Praise for *Art...*

"Mitchell explores some of AI's main domains: visual recognition, reinforcement learning and language processing. In each area, she explains the nuts and bolts, praises headline-grabbing breakthroughs, and then gives a reality check to those who might see human-like general intelligence in narrow exploits."
—Matthew Hutson, *The Wall Street Journal*

"A surprisingly lucid introduction to techniques that are making computers smarter."
—*Kirkus Reviews*

"Mitchell . . . ably illustrates the current state of artificial intelligence, debunking claims about computers that match or surpass human intelligence . . . Taking care to keep the text accessible, Mitchell lightens things with amusing facts, such as how *Star Trek*'s ship computer remains the gold standard for many AI researchers. This worthy volume should assuage lay readers' fears about AI, while also reassuring people drawn to the field that much work remains to be done."
—*Publishers Weekly*

"[*Artificial Intelligence*'s] historical grounding makes for a worthy and compelling narrative in itself. There are also ample contemporary topics explored in great detail, such as AI applications in image recognition, autonomous vehicles, voice recognition, and the impressive translation that today's popular search engines now provide."
—Jim Hahn, *Library Journal*

"An authoritative stroll through the development and state of play of [AI]."
—Philip Ball, *Prospect*

"In Mitchell's telling, artificial intelligence (AI) raises extraordinary issues that have disquieting implications for humanity . . . [Mitchell] provides readers with insightful, common-sense scrutiny of how these and related topics pervade the discipline of artificial intelligence."
—Howard Schneider, *Undark*

"A remarkably clear and readable primer."
—Brian Bergstein, *OneZero*

"[*Artificial Intelligence*] describ[es] wonderfully clearly how different AI applications actually work, and hence helps understand their strengths and limitations.

I would say these are the most illuminating simple yet meaningful explanations I've read of—for example—reinforcement learning, convolutional neural networks, word vectors, etc. I wish I'd had this book when I first started reading some of the AI literature." —Diane Coyle, *The Enlightened Economist*

"There are a lot of books about artificial intelligence . . . Melanie Mitchell's new book is . . . , in my opinion after surveying much of this literature, the most *intelligent* book on the subject." —Peter Kassan, *Skeptic*

"This book describes AI in clear and accessible terms. It cuts through the hype to present a sane assessment with no agenda apart from a desire to inform." —Stephen Few, *Visual Business Intelligence*

"Without a doubt, Mitchell's book sets a new standard in giving an understanding of what's possible and how difficult it is to go further. It should be read by every journalist, PR person and politician before they pump out yet more hype on the AI future." —*Popular Science* (UK)

"If you think you understand AI and all the related issues, you don't. By the time you finish this exceptionally lucid and riveting book, you will breathe more easily and more wisely."
—Michael S. Gazzaniga, director of the SAGE Center for the Study of the Mind at the University of California, Santa Barbara, and author of *The Consciousness Instinct*

"Melanie Mitchell's very intelligent, clear, and sensible book is a welcome corrective to the exaggerated fears and hopes for AI, and the perfect primer to start understanding how the systems actually work."
—Alison Gopnik, professor of psychology at the University of California, Berkeley, and author of *The Philosophical Baby: What Children's Minds Tell Us About Truth, Love, and the Meaning of Life*

"Melanie Mitchell's book is a must-read for anyone interested in the emerging revolution of AI, machine learning, and big data. This book can be, and should be, read by the proverbial man or woman in the street, Silicon Valley gurus, and members of Congress, as well as by professional scientists and engineers. They will all profit enormously from it."
—Geoffrey West, Shannan Distinguished Professor at the Santa Fe Institute and author of *Scale: The Universal Laws of Life, Growth, and Death in Organisms, Cities, and Companies*

"Melanie Mitchell takes us on an enlightening tour of how artificial intelligence currently works, and how it falls short of true human understanding. The challenges and opportunities discussed in this book will be crucial in shaping the future of humanity and technology."

—Sean Carroll, author of *Something Deeply Hidden:
Quantum Worlds and the Emergence of Spacetime*

"Melanie Mitchell writes about AI with a warm, friendly voice and an unpretentious brilliance that no machine could hope to match . . . for now."

—Steven Strogatz, professor of mathematics at Cornell University and
author of *Infinite Powers: How Calculus Reveals the Secrets of the Universe*

"Melanie Mitchell deftly provides the reader with a keen, clear-sighted account of the history of AI and neural networks. What is most impressive is that without getting too technical, Mitchell sketches enough details and includes enough clever illustrations that one gets a good intuitive understanding of AI, both its special-purpose machines and its attempts at developing a more general intelligence. A wonderfully informative book."

—John Allen Paulos, professor of mathematics at Temple University and
author of *Innumeracy: Mathematical Illiteracy and Its Consequences*

© Santa Fe Institute

MELANIE MITCHELL
Artificial Intelligence

Melanie Mitchell has a PhD in computer science from the University of Michigan, where she studied with the cognitive scientist and writer Douglas Hofstadter; together, they created the Copycat program, which makes creative analogies in an idealized world. The author and editor of six books and numerous scholarly papers, she is currently a professor of computer science at Portland State University and an external professor at the Santa Fe Institute.

Artificial Intelligence

Artificial Intelligence

·

A Guide for Thinking Humans

·

Melanie Mitchell

·

Picador

Farrar, Straus and Giroux

New York

Picador

120 Broadway, New York 10271

Copyright © 2019 by Melanie Mitchell
All rights reserved
Printed in the United States of America
Originally published in 2019 by Farrar, Straus and Giroux
First Picador paperback edition, 2020

The Library of Congress has cataloged the
Farrar, Straus and Giroux hardcover edition as follows:
Names: Mitchell, Melanie (Computer scientist), author.
Title: Artificial intelligence : a guide for thinking humans / Melanie Mitchell.
Description: New York : Farrar, Straus and Giroux, 2019. | Includes bibliographical
    references and index.
Identifiers: LCCN 2019011197 | ISBN 9780374257835 (hardcover)
Subjects: LCSH: Artificial intelligence. | Machine learning.
Classification: LCC Q335 .M58 2019 | DDC 006.3—dc23
LC record available at https://lccn.loc.gov/2019011197

Picador Paperback ISBN: 978-1-250-75804-0

Designed by Richard Oriolo

Our books may be purchased in bulk for promotional, educational, or
business use. Please contact your local bookseller or the Macmillan Corporate
and Premium Sales Department at 1-800-221-7945, extension 5442, or
by e-mail at MacmillanSpecialMarkets@macmillan.com.

Picador® is a U.S. registered trademark and is used by Macmillan Publishing Group, LLC,
under license from Pan Books Limited.

For book club information, please visit facebook.com/picadorbookclub or
e-mail marketing@picadorusa.com.

picadorusa.com • instagram.com/picador
twitter.com/picadorusa • facebook.com/picadorusa

9   10   8

To my parents,
who taught me how to be a thinking human, and so much more

Contents

Artificial Intelligence

Prologue:

▪

Terrified

▪

Computers seem to be getting smarter at an alarming rate, but one thing they still can't do is appreciate irony. That's what was on my mind a few years ago, when, on my way to a discussion about artificial intelligence (AI), I got lost in the capital of searching and finding—the Googleplex, Google's world headquarters in Mountain View, California. What's more, I was lost inside the Google Maps building. Irony squared.

The Maps building itself had been easy to find. A Google Street View car was parked by the front door, a hulking appendage crowned by a red-and-black soccer ball of a camera sticking up from its roof. However, once inside, with my prominent "Visitor" badge assigned by security, I wandered, embarrassed, among warrens of cubicles occupied by packs of Google workers, headphones over ears, intently typing on Apple desktops.

After some (map-less) random search, I finally found the conference room assigned for the daylong meeting and joined the group gathered there.

The meeting, in May 2014, had been organized by Blaise Agüera y Arcas, a young computer scientist who had recently left a top position at Microsoft to help lead Google's machine intelligence effort. Google started out in 1998 with one "product": a website that used a novel, extraordinarily successful method for searching the web. Over the years, Google has evolved into the world's most important tech company and now offers a vast array of products and services, including Gmail, Google Docs, Google Translate, YouTube, Android, many more that you might use every day, and some that you've likely never heard of.

Google's founders, Larry Page and Sergey Brin, have long been motivated by the idea of creating artificial intelligence in computers, and this quest has become a major focus at Google. In the last decade, the company has hired a profusion of AI experts, most notably Ray Kurzweil, a well-known inventor and a controversial futurist who promotes the idea of an AI Singularity, a time in the near future when computers will become smarter than humans. Google hired Kurzweil to help realize this vision. In 2011, Google created an internal AI research group called Google Brain; since then, the company has also acquired an impressive array of AI start-up companies with equally optimistic names: Applied Semantics, Deep-Mind, and Vision Factory, among others.

In short, Google is no longer merely a web-search portal—not by a long shot. It is rapidly becoming an applied AI company. AI is the glue that unifies the diverse products, services, and blue-sky research efforts offered by Google and its parent company, Alphabet. The company's ultimate aspiration is reflected in the original mission statement of its DeepMind group: "Solve intelligence and use it to solve everything else."[1]

AI and *GEB*

I was pretty excited to attend an AI meeting at Google. I had been working on various aspects of AI since graduate school in the 1980s and had been tremendously impressed by what Google had accomplished. I also thought I had some good ideas to contribute. But I have to admit that I was there only as a tagalong. The meeting was happening so that a group of select Google AI researchers could hear from and converse with Douglas Hofstadter, a legend in AI and the author of a famous book cryptically titled *Gödel, Escher, Bach: an Eternal Golden Braid*, or more succinctly, *GEB* (pronounced "gee-ee-bee"). If you're a computer scientist, or a computer enthusiast, it's likely you've heard of it, or read it, or tried to read it.

Written in the 1970s, *GEB* was an outpouring of Hofstadter's many intellectual passions—mathematics, art, music, language, humor, and wordplay, all brought together to address the deep questions of how intelligence, consciousness, and the sense of self-awareness that each human experiences so fundamentally can emerge from the non-intelligent, non-conscious substrate of biological cells. It's also about how intelligence and self-awareness might eventually be attained by computers. It's a unique book; I don't know of any other book remotely like it. It's not an easy read, and yet it became a bestseller and won both the Pulitzer Prize and the National Book Award. Without a doubt, *GEB* inspired more young people to pursue AI than any other book. I was one of those young people.

In the early 1980s, after graduating from college with a math degree, I was living in New York City, teaching math in a prep school, unhappy, and casting about for what I really wanted to do in life. I discovered *GEB* after reading a rave review in *Scientific American*. I went out and bought the book immediately. Over the next several weeks, I devoured it, becoming increasingly convinced that not only did I want to become an AI researcher but I specifically wanted to work with Douglas Hofstadter. I had never before felt so strongly about a book, or a career choice.

At the time, Hofstadter was a professor in computer science at Indiana University, and my quixotic plan was to apply to the computer science PhD program there, arrive, and then persuade Hofstadter to accept me as a student. One minor problem was that I had never taken even one computer science course. I had grown up with computers; my father was a hardware engineer at a 1960s tech start-up company, and as a hobby he built a mainframe computer in our family's den. The refrigerator-sized Sigma 2 machine wore a magnetic button proclaiming "I pray in FORTRAN," and as a child I was half-convinced it did, quietly at night, while the rest of the family was asleep. Growing up in the 1960s and '70s, I learned a bit of each of the popular languages of the day: FORTRAN, then BASIC, then Pascal, but I knew next to nothing about proper programming techniques, not to mention anything else an incoming computer science graduate student needs to know.

To speed along my plan, I quit my teaching job at the end of the school year, moved to Boston, and started taking introductory computer science courses to prepare for my new career. A few months into my new life, I was on the campus of the Massachusetts Institute of Technology, waiting for a class to begin, and I caught sight of a poster advertising a lecture by Douglas Hofstadter, to take place in two days on that very campus. I did a double take; I couldn't believe my good fortune. I went to the lecture, and after a long wait for my turn in a crowd of admirers I managed to speak to Hofstadter. It turned out he was in the middle of a yearlong sabbatical at MIT, after which he was moving from Indiana to the University of Michigan in Ann Arbor.

To make a long story short, after some persistent pursuit on my part, I persuaded Hofstadter to take me on as a research assistant, first for a summer, and then for the next six years as a graduate student, after which I graduated with a doctorate in computer science from Michigan. Hofstadter and I have kept in close touch over the years and have had many discussions about AI. He knew of my interest in Google's AI research and was nice enough to invite me to accompany him to the Google meeting.

Chess and the First Seed of Doubt

The group in the hard-to-locate conference room consisted of about twenty Google engineers (plus Douglas Hofstadter and myself), all of whom were members of various Google AI teams. The meeting started with the usual going around the room and having people introduce themselves. Several noted that their own careers in AI had been spurred by reading GEB at a young age. They were all excited and curious to hear what the legendary Hofstadter would say about AI. Then Hofstadter got up to speak. "I have some remarks about AI research in general, and here at Google in particular." His voice became passionate. "I am terrified. Terrified."

Hofstadter went on.[2] He described how, when he first started working on AI in the 1970s, it was an exciting prospect but seemed so far from being realized that there was no "danger on the horizon, no sense of it actually *happening*." Creating machines with humanlike intelligence was a profound intellectual adventure, a long-term research project whose fruition, it had been said, lay at least "one hundred Nobel prizes away."[3] Hofstadter believed AI was possible in principle: "The 'enemy' were people like John Searle, Hubert Dreyfus, and other skeptics, who were saying it was impossible. They did not understand that a brain is a hunk of matter that obeys physical law and the computer can simulate anything . . . the level of neurons, neurotransmitters, et cetera. In theory, it can be done." Indeed, Hofstadter's ideas about simulating intelligence at various levels— from neurons to consciousness—were discussed at length in *GEB* and had been the focus of his own research for decades. But in practice, until recently, it seemed to Hofstadter that general "human-level" AI had no chance of occurring in his (or even his children's) lifetime, so he didn't worry much about it.

Near the end of *GEB*, Hofstadter had listed "Ten Questions and Speculations" about artificial intelligence. Here's one of them: "Will there be chess programs that can beat anyone?" Hofstadter's speculation was "no." "There may be programs which can beat anyone at chess, but they will not be exclusively chess players. They will be programs of *general* intelligence."[4]

At the Google meeting in 2014, Hofstadter admitted that he had been "dead wrong." The rapid improvement in chess programs in the 1980s and '90s had sown the first seed of doubt in his appraisal of AI's short-term prospects. Although the AI pioneer Herbert Simon had predicted in 1957 that a chess program would be world champion "within 10 years," by the mid-1970s, when Hofstadter was writing *GEB*, the best computer chess programs played only at the level of a good (but not great) amateur. Hofstadter had befriended Eliot Hearst, a chess champion and psychology professor who had written extensively on how human chess experts differ from computer chess programs. Experiments showed that expert human players rely on quick recognition of patterns on the chessboard to decide on a move rather than the extensive brute-force look-ahead search that all chess programs use. During a game, the best human players can perceive a configuration of pieces as a particular "kind of position" that requires a certain "kind of strategy." That is, these players can quickly recognize particular configurations and strategies as instances of higher-level concepts. Hearst argued that without such a general ability to perceive patterns and recognize abstract concepts, chess programs would never reach the level of the best humans. Hofstadter was persuaded by Hearst's arguments.

However, in the 1980s and '90s, computer chess saw a big jump in improvement, mostly due to the steep increase in computer speed. The best programs still played in a very unhuman way: performing extensive look-ahead to decide on the next move. By the mid-1990s, IBM's Deep Blue machine, with specialized hardware for playing chess, had reached the Grandmaster level, and in 1997 the program defeated the reigning world chess champion, Garry Kasparov, in a six-game match. Chess mastery, once seen as a pinnacle of human intelligence, had succumbed to a brute-force approach.

Music: The Bastion of Humanity

Although Deep Blue's win generated a lot of hand-wringing in the press about the rise of intelligent machines, "true" AI still seemed quite distant. Deep Blue could play chess, but it couldn't do anything else. Hofstadter had been wrong about chess, but he still stood by the other speculations in *GEB*, especially the one he had listed first:

QUESTION: Will a computer ever write beautiful music?
SPECULATION: Yes but not soon.

Hofstadter continued,

Music is a language of emotions, and until programs have emotions as complex as ours, there is no way a program will write anything beautiful. There can be "forgeries"—shallow imitations of the syntax of earlier music—but despite what one might think at first, there is much more to musical expression than can be captured in syntactic rules. . . . To think . . . that we might soon be able to command a preprogrammed mass-produced mail-order twenty-dollar desk-model "music box" to bring forth from its sterile circuitry pieces which Chopin or Bach might have written had they lived longer is a grotesque and shameful misestimation of the depth of the human spirit.[5]

Hofstadter described this speculation as "one of the most important parts of *GEB*—I would have staked my life on it."

In the mid-1990s, Hofstadter's confidence in his assessment of AI was again shaken, this time quite profoundly, when he encountered a program written by a musician, David Cope. The program was called Experiments in Musical Intelligence, or EMI (pronounced "Emmy"). Cope, a composer and music professor, had originally developed EMI to aid him in his own composing process by automatically creating pieces in Cope's specific style. However, EMI became famous for creating pieces in the style of classical composers such as Bach and Chopin. EMI composes by following a large set of rules, developed by Cope, that are meant to capture a general syntax of composition. These rules are applied to copious examples from a particular composer's opus in order to produce a new piece "in the style" of that composer.

Back at our Google meeting, Hofstadter spoke with extraordinary emotion about his encounters with EMI:

> I sat down at my piano and I played one of EMI's mazurkas "in the style of Chopin." It didn't sound exactly like Chopin, but it sounded enough like Chopin, and like coherent music, that I just felt *deeply* troubled.
>
> Ever since I was a child, music has thrilled me and moved me to the very core. And every piece that I love feels like it's a direct message from the emotional heart of the human being who composed it. It feels like it is giving me access to their innermost soul. And it feels like there is *nothing* more human in the world than that expression of music. Nothing. The idea that pattern manipulation of the most superficial sort can yield things that sound as if they are coming from a human being's heart is very, very troubling. I was just completely thrown by this.

Hofstadter then recounted a lecture he gave at the prestigious Eastman School of Music, in Rochester, New York. After describing EMI, Hofstadter had asked the Eastman audience—including several music theory and composition faculty—to guess which of two pieces a pianist played for them was a (little-known) mazurka by Chopin and which had been composed by EMI. As one audience member described later, "The first mazurka had grace and charm, but not 'true-Chopin' degrees of invention and large-scale fluidity . . . The second was clearly the genuine Chopin, with a lyrical melody; large-scale, graceful chromatic modulations; and a natural, balanced form."[6] Many of the faculty agreed and, to Hofstadter's shock, voted EMI for the first piece and "real-Chopin" for the second piece. The correct answers were the reverse.

In the Google conference room, Hofstadter paused, peering into our faces. No one said a word. At last he went on. "I was terrified by EMI. Terrified. I hated it, and was extremely threatened by it. It was threatening to destroy what I most cherished about humanity. I think EMI was the most quintessential example of the fears that I have about artificial intelligence."

Google and the Singularity

Hofstadter then spoke of his deep ambivalence about what Google itself was trying to accomplish in AI—self-driving cars, speech recognition, natural-language understanding, translation between languages, computer-generated art, music composition, and more. Hofstadter's worries were underlined by Google's embrace of Ray Kurzweil and his vision of the Singularity, in which AI, empowered by its ability to improve itself and learn on its own, will quickly reach, and then exceed, human-level intelligence. Google, it seemed, was doing everything it could to accelerate that vision. While Hofstadter strongly doubted the premise of the Singularity, he admitted that Kurzweil's predictions still disturbed him. "I was terrified by the scenarios. Very skeptical, but at the same time, I thought, maybe their timescale is off, but maybe they're right. We'll be completely caught off guard. We'll think nothing is happening and all of a sudden, before we know it, computers will be smarter than us."

If this actually happens, "we will be superseded. We will be relics. We will be left in the dust.

"Maybe this is going to happen, but I don't want it to happen *soon*. I don't want my children to be left in the dust."

Hofstadter ended his talk with a direct reference to the very Google engineers in that room, all listening intently: "I find it very scary, very troubling, very sad, and I find it terrible, horrifying, bizarre, baffling, bewildering, that people are rushing ahead blindly and deliriously in creating these things."

Why Is Hofstadter Terrified?

I looked around the room. The audience appeared mystified, embarrassed even. To these Google AI researchers, none of this was the least bit terrifying. In fact, it was old news. When Deep Blue beat Kasparov, when EMI started composing Chopin-like mazurkas, and when Kurzweil wrote his first book on the Singularity, many of these engineers had been in high school, probably reading GEB and loving it, even though its AI prognostications were a bit out of date. The reason they were working at Google was precisely to make AI happen—not in a hundred years, but now, as soon as possible. They didn't understand what Hofstadter was so stressed out about.

People who work in AI are used to encountering the fears of people outside the field, who have presumably been influenced by the many science fiction movies depicting superintelligent machines that turn evil. AI researchers are also familiar with the worries that increasingly sophisticated AI will replace humans in some jobs, that AI applied to big data sets could subvert privacy and enable subtle discrimination, and that ill-understood AI systems allowed to make autonomous decisions have the potential to cause havoc.

Hofstadter's terror was in response to something entirely different. It was not about AI becoming too smart, too invasive, too malicious, or even too useful. Instead, he was terrified that intelligence, creativity, emotions, and maybe even consciousness itself would be too *easy* to produce—that what he valued most in humanity would end up being nothing more than a "bag of tricks," that a superficial set of brute-force algorithms could explain the human spirit.

As *GEB* made abundantly clear, Hofstadter firmly believes that the mind and all its characteristics emerge wholly from the physical substrate of the brain and the rest of the body, along with the body's interaction with the physical world. There is nothing immaterial or incorporeal lurking there. The issue that worries him is really one of complexity. He fears that AI might show us that the human qualities we most value are disappointingly simple to mechanize. As Hofstadter explained to me after the meeting, here referring to Chopin, Bach, and other paragons of humanity, "If such minds of infinite subtlety and complexity and emotional depth could be trivialized by a small chip, it would destroy my sense of what humanity is about."

I Am Confused

Following Hofstadter's remarks, there was a short discussion, in which the nonplussed audience prodded Hofstadter to further explain his fears about AI and about Google in particular. But a communication barrier remained. The meeting continued, with project presentations, group discussion, coffee breaks, the usual—none of it really touching on Hofstadter's comments. Close to the end of the meeting, Hofstadter asked the participants for their thoughts about the near-term future of AI. Several of the Google researchers predicted that general human-level AI would likely emerge within the next thirty years, in large part due to Google's own advances on the brain-inspired method of "deep learning."

I left the meeting scratching my head in confusion. I knew that Hofstadter had been troubled by some of Kurzweil's Singularity writings, but I had never before appreciated the degree of his emotion and anxiety. I also had known that Google was pushing hard on AI research, but I was startled by the optimism several people there expressed about how soon AI would reach a general "human" level. My own view had been that AI had progressed a lot in some narrow areas but was still nowhere close to having the broad, general intelligence of humans, and it would not get there in a century, let alone thirty years. And I had thought that people who believed otherwise were vastly underestimating the complexity of human intelligence. I had read Kurzweil's books and had found them largely ridiculous. However, listening to all the comments at the meeting, from people I respected and admired, forced me to critically examine my own views. While assuming that these AI researchers underestimated humans, had I in turn underestimated the power and promise of current-day AI?

Over the months that followed, I started paying more attention to the discussion surrounding these questions. I started to notice the slew of articles, blog posts, and entire books by prominent people suddenly telling us we should start worrying, right now, about the perils of "superhuman" AI. In 2014, the physicist Stephen Hawking proclaimed, "The development of full artificial intelligence could spell the end of the human race."[7] In the same year, the entrepreneur Elon Musk, founder of the Tesla and SpaceX companies, said that artificial intelligence is probably "our biggest existential threat" and that "with artificial intelligence we are summoning the demon."[8] Microsoft's cofounder Bill Gates concurred: "I agree with Elon Musk and some others on this and don't understand why some people are not concerned."[9] The philosopher Nick Bostrom's book *Superintelligence*, on the potential dangers of machines becoming smarter than humans, became a surprise bestseller, despite its dry and ponderous style.

Other prominent thinkers were pushing back. Yes, they said, we should make sure that AI programs are safe and don't risk harming humans, but any reports of near-term superhuman AI are greatly exaggerated. The entrepreneur and activist Mitchell Kapor advised, "Human intelligence is a marvelous, subtle, and poorly understood phenomenon. There is no danger of duplicating it anytime soon."[10] The roboticist (and former director of MIT's AI Lab) Rodney Brooks agreed, stating that we "grossly overestimate the capabilities of machines—those of today and of the next few decades."[11] The psychologist and AI researcher Gary Marcus went so far as to assert that in the quest to create "strong AI"—that is, *general* human-level AI—"there has been almost no progress."[12]

I could go on and on with dueling quotations. In short, what I found is that the field of AI is in turmoil. Either a huge amount of progress has been made, or almost none at all. Either we are within spitting distance of "true" AI, or it is centuries away. AI will solve all our problems, put us all out of a job, destroy the human race, or cheapen our humanity. It's either a noble quest or "summoning the demon."

What This Book Is About

This book arose from my attempt to understand the true state of affairs in artificial intelligence—what computers can do now, and what we can expect from them over the next decades. Hofstadter's provocative comments at the Google meeting were something of a wake-up call for me, as were the Google researchers' confident responses about AI's near-term future. In the chapters that follow, I try to sort out how far artificial intelligence has come, as well as elucidate its disparate—and sometimes conflicting—goals. In doing so, I consider how some of the most prominent AI systems actually work, and investigate how successful they are and where their limitations lie. I look at the extent to which computers can now do things that we believe to require high levels of intelligence—beating humans at the most intellectually demanding games, translating between languages, answering complex questions, navigating vehicles in challenging terrain. And I examine how they fare at the things we take for granted, the everyday tasks we humans perform without conscious thought: recognizing faces and objects in images, understanding spoken language and written text, and using the most basic common sense.

I also try to make sense of the broader questions that have fueled debates about AI since its inception: What do we actually mean by "general human" or even "superhuman" intelligence? Is current AI close to this level, or even on a trajectory to get there? What are the dangers? What aspects of our intelligence do we most cherish, and to what extent would human-level AI challenge how we think about our own humanness? To use Hofstadter's terms, how terrified should we be?

This book is not a general survey or history of artificial intelligence. Rather, it is an in-depth exploration of some of the AI methods that probably affect your life, or will soon, as well as the AI efforts that perhaps go furthest in challenging our sense of human uniqueness. My aim is for you to share in my own exploration and, like me, to come away with a clearer sense of what the field has accomplished and how much further there is to go before our machines can argue for their own humanity.

Part I

·

Background

·

1

The Roots of

Artificial Intelligence

Two Months and Ten Men at Dartmouth

The dream of creating an intelligent machine—one that is as smart as or smarter than humans—is centuries old but became part of modern science with the rise of digital computers. In fact, the ideas that led to the first programmable computers came out of mathematicians' attempts to understand human thought—particularly logic—as a mechanical process of "symbol manipulation." Digital computers are essentially symbol manipulators, pushing around combinations of the symbols 0 and 1. To pioneers of computing like Alan Turing and John von Neumann, there were strong analogies between computers and the human brain, and it seemed obvious to them that human intelligence could be replicated in computer programs.

Most people in artificial intelligence trace the field's official founding to a small workshop in 1956 at Dartmouth College organized by a young mathematician named John McCarthy.

In 1955, McCarthy, aged twenty-eight, joined the mathematics faculty at Dartmouth. As an undergraduate, he had learned a bit about both psychology and the nascent field of "automata theory" (later to become computer science) and had become intrigued with the idea of creating a thinking machine. In graduate school in the mathematics department at Princeton, McCarthy had met a fellow student, Marvin Minsky, who shared his fascination with the potential of intelligent computers. After graduating, McCarthy had short-lived stints at Bell Labs and IBM, where he collaborated, respectively, with Claude Shannon, the inventor of information theory, and Nathaniel Rochester, a pioneering electrical engineer. Once at Dartmouth, McCarthy persuaded Minsky, Shannon, and Rochester to help him organize "a 2 month, 10 man study of artificial intelligence to be carried out during the summer of 1956."[1] The term *artificial intelligence* was McCarthy's invention; he wanted to distinguish this field from a related effort called cybernetics.[2] McCarthy later admitted that no one really liked the name—after all, the goal was *genuine*, not "artificial," intelligence—but "I had to call it something, so I called it 'Artificial Intelligence.'"[3]

The four organizers submitted a proposal to the Rockefeller Foundation asking for funding for the summer workshop. The proposed study was, they wrote, based on "the conjecture that every aspect of learning or any other feature of intelligence can be in principle so precisely described that a machine can be made to simulate it."[4] The proposal listed a set of topics to be discussed—natural-language processing, neural networks, machine learning, abstract concepts and reasoning, creativity—that have continued to define the field to the present day.

Even though the most advanced computers in 1956 were about a million times slower than today's smartphones, McCarthy and colleagues were optimistic that AI was in close reach: "We think that a significant advance can be made in one or more of these problems if a carefully selected group of scientists work on it together for a summer."[5]

Obstacles soon arose that would be familiar to anyone organizing a scientific workshop today. The Rockefeller Foundation came through with only half the requested amount of funding. And it turned out to be harder than McCarthy had thought to persuade the participants to actually come and then stay, not to mention agree on anything. There were lots of interesting discussions but not a lot of coherence. As usual in such meetings, "Everyone had a different idea, a hearty ego, and much enthusiasm for their own plan."[6] However, the Dartmouth summer of AI did produce a few very important outcomes. The field itself was named, and its general goals were outlined. The soon-to-be "big four" pioneers of the field—McCarthy, Minsky, Allen Newell, and Herbert Simon—met and did some planning for the future. And for whatever reason, these four came out of the meeting with tremendous optimism for the field. In the early 1960s, McCarthy founded the Stanford Artificial Intelligence Project, with the "goal of building a fully intelligent machine in a decade."[7] Around the same time, the future Nobel laureate Herbert Simon predicted, "Machines will be capable, within twenty years, of doing any work that a man can do."[8] Soon after, Marvin Minsky, founder of the MIT AI Lab, forecast that "within a generation . . . the problems of creating 'artificial intelligence' will be substantially solved."[9]

Definitions, and Getting On with It

None of these predicted events have yet come to pass. So how far do we remain from the goal of building a "fully intelligent machine"? Would such a machine require us to reverse engineer the human brain in all its complexity, or is there a shortcut, a clever set of yet-unknown algorithms, that can produce what we recognize as full intelligence? What does "full intelligence" even mean?

"Define your terms . . . or we shall never understand one another."[10] This admonition from the eighteenth-century philosopher Voltaire is a challenge for anyone talking about artificial intelligence, because its central notion—intelligence—remains so ill-defined. Marvin Minsky himself coined the phrase "suitcase word"[11] for terms like *intelligence* and its many cousins, such as *thinking, cognition, consciousness,* and *emotion.* Each is packed like a suitcase with a jumble of different meanings. *Artificial intelligence* inherits this packing problem, sporting different meanings in different contexts.

Most people would agree that humans are intelligent and specks of dust are not. Likewise, we generally believe that humans are more intelligent than worms. As for human intelligence, IQ is measured on a single scale, but we also talk about the different dimensions of intelligence: emotional, verbal, spatial, logical, artistic, social, and so forth. Thus, intelligence can be binary (something is or is not intelligent), on a continuum (one thing is more intelligent than another thing), or multidimensional (someone can have high verbal intelligence but low emotional intelligence). Indeed, the word *intelligence* is an over-packed suitcase, zipper on the verge of breaking.

For better or worse, the field of AI has largely ignored these various distinctions. Instead, it has focused on two efforts: one scientific and one practical. On the scientific side, AI researchers are investigating the mechanisms of "natural" (that is, biological) intelligence by trying to embed it in computers. On the practical side, AI proponents simply want to create computer programs that perform tasks as well as or better than humans, without worrying about whether these programs are actually *thinking* in the way humans think. When asked if their motivations are practical or scientific, many AI people joke that it depends on where their funding currently comes from.

In a recent report on the current state of AI, a committee of prominent researchers defined the field as "a branch of computer science that studies the properties of intelligence by synthesizing intelligence."[12] A bit circular, yes. But the same committee also admitted that it's hard to define the field, and that may be a good thing: "The lack of a precise, universally accepted definition of AI probably has helped the field to grow, blossom, and advance at an ever-accelerating pace."[13] Furthermore, the committee notes, "Practitioners, researchers, and developers of AI are instead guided by a rough sense of direction and an imperative to 'get on with it.'"

An Anarchy of Methods

At the 1956 Dartmouth workshop, different participants espoused divergent opinions about the correct approach to take to develop AI. Some people—generally mathematicians—promoted mathematical logic and deductive reasoning as the language of rational thought. Others championed inductive methods in which programs extract statistics from data and use probabilities to deal with uncertainty. Still others believed firmly in taking inspiration from biology and psychology to create brain-like programs. What you may find surprising is that the arguments among proponents of these various approaches persist to this day. And each approach has generated its own panoply of principles and techniques, fortified by specialty conferences and journals, with little communication among the subspecialties. A recent AI survey paper summed it up: "Because we don't deeply understand intelligence or know how to produce general AI, rather than cutting off any avenues of exploration, to truly make progress we should embrace AI's 'anarchy of methods.'"[14]

But since the 2010s, one family of AI methods—collectively called deep learning (or deep neural networks)—has risen above the anarchy to become the dominant AI paradigm. In fact, in much of the popular media, the term *artificial intelligence* itself has come to mean "deep learning." This is an unfortunate inaccuracy, and I need to clarify the distinction. AI is a field that includes a broad set of approaches, with the goal of creating machines with intelligence. Deep learning is only one such approach. Deep learning is itself one method among many in the field of *machine learning*, a subfield of AI in which machines "learn" from data or from their own "experiences." To better understand these various distinctions, it's important to understand a philosophical split that occurred early in the AI research community: the split between so-called symbolic and subsymbolic AI.

Symbolic AI

First let's look at *symbolic AI*. A symbolic AI program's knowledge consists of words or phrases (the "symbols"), typically understandable to a human, along with rules by which the program can combine and process these symbols in order to perform its assigned task.

I'll give you an example. One early AI program was confidently called the General Problem Solver,[15] or GPS for short. (Sorry about the confusing acronym; the General Problem Solver predated the Global Positioning System.) GPS could solve problems such as the "Missionaries and Cannibals" puzzle, which you might have tackled yourself as a child. In this well-known conundrum, three missionaries and three cannibals all need to cross a river, but their boat holds only two people. If at any time the (hungry) cannibals outnumber the (tasty-looking) missionaries on one side of the river . . . well, you probably know what happens. How do all six get across the river intact?

The creators of the General Problem Solver, the cognitive scientists Herbert Simon and Allen Newell, had recorded several students "thinking out loud" while solving this and other logic puzzles. Simon and Newell then designed their program to mimic what they believed were the students' thought processes.

I won't go into the details of how GPS worked, but its symbolic nature can be seen by the way the program's instructions were encoded. To set up the problem, a human would write code for GPS that looked something like this:

```
CURRENT STATE:
LEFT-BANK = [3 MISSIONARIES, 3 CANNIBALS, 1 BOAT]
RIGHT-BANK = [EMPTY]
```

```
DESIRED STATE:
LEFT-BANK = [EMPTY]
RIGHT-BANK = [3 MISSIONARIES, 3 CANNIBALS, 1 BOAT]
```

In English, these lines represent the fact that initially the left bank of the river "contains" three missionaries, three cannibals, and one boat, whereas the right bank doesn't contain any of these. The desired state represents the goal of the program—get everyone to the right bank of the river.

At each step in its procedure, GPS attempts to change its current state to make it more similar to the desired state. In its code, the program has "operators" (in the form of subprograms) that can transform the current state into a new state and "rules" that encode the constraints of the task.

For example, there is an operator that moves some number of missionaries and cannibals from one side of the river to the other:

MOVE (#MISSIONARIES, #CANNIBALS, FROM-SIDE, TO-SIDE)

The words inside the parentheses are called arguments, and when the program runs, it replaces these words with numbers or other words. That is, #MISSIONARIES is replaced with the number of missionaries to move, #CANNIBALS with the number of cannibals to move, and FROM-SIDE and TO-SIDE are replaced with "LEFT-BANK" or "RIGHT-BANK," depending on which riverbank the missionaries and cannibals are to be moved from. Encoded into the program is the knowledge that the boat is moved along with the missionaries and cannibals.

Before being able to apply this operator with specific values replacing the arguments, the program must check its encoded rules; for example, the maximum number of people that can move at a time is two, and the operator cannot be used if it will result in cannibals outnumbering missionaries on a riverbank.

While these symbols represent human-interpretable concepts such as *missionaries*, *cannibals*, *boat*, and *left bank*, the computer running this program of course has no knowledge of the meaning of these symbols. You could replace all occurrences of "MISSIONARIES" with "Z372B" or any other nonsense string, and the program would work in exactly the same way. This is part of what the term *General* refers to in General Problem Solver. To the computer, the "meaning" of the symbols derives from the ways in which they can be combined, related to one another, and operated on.

Advocates of the symbolic approach to AI argued that to attain intelligence in computers, it would not be necessary to build programs that mimic the brain. Instead, the argument goes, general intelligence can be captured entirely by the right kind of symbol-processing program. Agreed, the workings of such a program would be vastly more complex than the Missionaries and Cannibals example, but it would still consist of symbols, combinations of symbols, and rules and operations on symbols. Symbolic AI of the kind illustrated by GPS ended up dominating the field for its first three decades, most notably in the form of *expert systems*, in which human experts devised rules for computer programs to use in tasks such as medical diagnosis and legal decision-making. There are several active branches of AI that still employ symbolic AI; I'll describe examples of it later, particularly in discussions of AI approaches to reasoning and common sense.

Subsymbolic AI: Perceptrons

Symbolic AI was originally inspired by mathematical logic as well as by the way people described their conscious thought processes. In contrast, *subsymbolic* approaches to AI took inspiration from neuroscience and sought to capture the sometimes-*unconscious* thought processes underlying what some have called fast perception, such as recognizing faces or identifying spoken words. Subsymbolic AI programs do not contain the kind of human-understandable language we saw in the Missionaries and Cannibals example above. Instead, a subsymbolic program is essentially a stack of equations—a thicket of often hard-to-interpret operations on numbers. As I'll explain shortly, such systems are designed to learn from data how to perform a task.

An early example of a subsymbolic, brain-inspired AI program was the perceptron, invented in the late 1950s by the psychologist Frank Rosenblatt.[16] The term *perceptron* may sound a bit 1950s science-fiction-y to our modern ears (as we'll see, it was soon followed by the "cognitron" and the "neocognitron"), but the perceptron was an important milestone in AI and was the influential great-grandparent of modern AI's most successful tool, deep neural networks.

Rosenblatt's invention of perceptrons was inspired by the way in which neurons process information. A neuron is a cell in the brain that receives electrical or chemical input from other neurons that connect to it. Roughly speaking, a neuron sums up all the inputs it receives from other neurons, and if the total sum reaches a certain threshold level, the neuron fires. Importantly, different connections (*synapses*) from other neurons to a given neuron have different strengths; in calculating the sum of its inputs, the given neuron gives more weight to inputs from stronger connections than inputs from weaker connections. Neuroscientists believe that adjust-

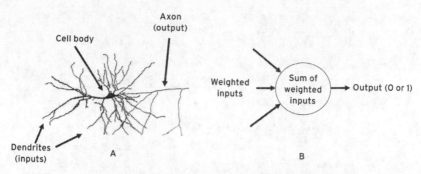

FIGURE 1: *A*, a neuron in the brain; *B*, a simple perceptron

ments to the strength of connections between neurons is a key part of how learning takes place in the brain.

To a computer scientist (or, in Rosenblatt's case, a psychologist), information processing in neurons can be simulated by a computer program—a perceptron—that has multiple numerical inputs and one output. The analogy between a neuron and a perceptron is illustrated in figure 1. Figure 1*A* shows a neuron, with its branching dendrites (fibers that carry inputs to the cell), cell body, and axon (that is, output channel) labeled. Figure 1*B* shows a simple perceptron. Analogous to the neuron, the perceptron adds up its inputs, and if the resulting sum is equal to or greater than the perceptron's *threshold*, the perceptron outputs the value 1 (it "fires"); otherwise it outputs the value 0 (it "does not fire"). To simulate the different strengths of connections to a neuron, Rosenblatt proposed that a numerical *weight* be assigned to each of a perceptron's inputs; each input is multiplied by its weight before being added to the sum. A perceptron's *threshold* is simply a number set by the programmer (or, as we'll see, learned by the perceptron itself).

In short, a perceptron is a simple program that makes a yes-or-no (1 or 0) decision based on whether the sum of its weighted inputs meets a threshold value. You probably make some decisions like this in your life. For example, you might get input from several friends on how much they liked a particular movie, but you trust some of those friends' taste in movies more than others. If the total amount of "friend enthusiasm"—giving

FIGURE 2: Examples of handwritten digits

more weight to your more trusted friends—is high enough (that is, greater than some unconscious threshold), you decide to go to the movie. This is how a perceptron would decide about movies, if only it had friends.

Inspired by networks of neurons in the brain, Rosenblatt proposed that networks of perceptrons could perform visual tasks such as recognizing faces and objects. To get a flavor of how that might work, let's explore how a perceptron might be used for a particular visual task: recognizing handwritten digits like those in figure 2.

In particular, let's design a perceptron to be an 8 detector—that is, to output a 1 if its inputs are from an image depicting an 8, and to output a 0 if the image depicts some other digit. Designing such a detector requires us to (1) figure out how to turn an image into a set of numerical inputs, and (2) determine numbers to use for the perceptron's weights and threshold, so that it will give the correct output (1 for 8s, 0 for other digits). I'll go into some detail here because many of the same ideas will arise later in my discussions of neural networks and their applications in computer vision.

Our Perceptron's Inputs

Figure 3A shows an enlarged handwritten 8. Each grid square is a pixel with a numerical "intensity" value: white squares have an intensity of 0,

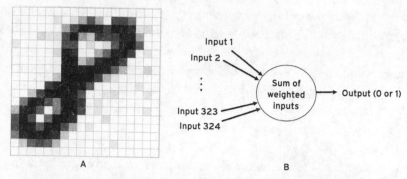

FIGURE 3: An illustration of a perceptron that recognizes handwritten 8s. Each pixel in the 18×18–pixel image corresponds to an input for the perceptron, yielding 324 (= 18×18) inputs.

black squares have an intensity of 1, and gray squares are in between. Let's assume that the images we give to our perceptron have been adjusted to be the same size as this one: 18 × 18 pixels. Figure 3B illustrates a perceptron for recognizing 8s. This perceptron has 324 (that is, 18 × 18) inputs, each of which corresponds to one of the pixels in the 18 × 18 grid. Given an image like the one in figure 3A, each of the perceptron's inputs is set to the corresponding pixel's intensity. Each of the inputs would have its own weight value (not shown in the figure).

Learning the Perceptron's Weights and Threshold

Unlike the symbolic General Problem Solver system that I described earlier, a perceptron doesn't have any explicit rules for performing its task; all of its "knowledge" is encoded in the numbers making up its weights and threshold. In his various papers, Rosenblatt showed that given the correct weight and threshold values, a perceptron like the one in figure 3B can perform fairly well on perceptual tasks such as recognizing simple handwritten digits. But how, exactly, can we determine the correct weights and threshold for a given task? Again, Rosenblatt proposed a brain-inspired answer: the perceptron should *learn* these values on its own. And how is it supposed to learn the correct values? Like the behavioral psychology theories popular at the time, Rosenblatt's idea was that perceptrons should learn via *conditioning*. Inspired in part by the behaviorist psychologist B. F. Skinner, who trained rats and pigeons to perform tasks by giving them positive and negative reinforcement, Rosenblatt's idea was that the perceptron should similarly be *trained* on examples: it should be rewarded when it fires correctly and punished when it errs. This form of conditioning is now known in AI as supervised learning. During training, the learning system is given an example, it produces an output, and it is then given a "supervision signal," which tells how much the system's output differs from the correct output. The system then uses this signal to adjust its weights and threshold.

The concept of supervised learning is a key part of modern AI, so it's worth discussing in more detail. Supervised learning typically requires a large set of positive examples (for instance, a collection of 8s written by different people) and negative examples (for instance, a collection of other handwritten digits, not including 8s). Each example is *labeled* by a human with its category—here, 8 or not-8. This label will be used as the supervision signal. Some of the positive and negative examples are used to *train* the system; these are called the *training set*. The remainder—the *test set*— is used to evaluate the system's performance after it has been trained, to see how well it has learned to answer correctly in general, not just on the training examples.

Perhaps the most important term in computer science is *algorithm*, which refers to a "recipe" of steps a computer can take in order to solve a particular problem. Frank Rosenblatt's primary contribution to AI was his design of a specific algorithm, called the perceptron-learning algorithm, by which a perceptron could be trained from examples to determine the weights and threshold that would produce correct answers. Here's how it works: Initially, the weights and threshold are set to random values between −1 and 1. In our example, the weight on the first input might be set to 0.2, the weight on the second input set to −0.6, and so on, and the threshold set to 0.7. A computer program called a random-number generator can easily generate these initial values.

Now we can start the training process. The first training example is given to the perceptron; at this point, the perceptron doesn't see the correct category label. The perceptron multiplies each input by its weight, sums up all the results, compares the sum with the threshold, and outputs either 1 or 0. Here, the output 1 means a guess of 8, and the output 0 means a guess of not-8. Now, the training process compares the perceptron's output with the correct answer given by the human-provided label (that is, 8 or not-8). If the perceptron is correct, the weights and threshold don't change. But if the perceptron is wrong, the weights and threshold are changed a little bit, making the perceptron's sum on this training example closer to producing the right answer. Moreover, the amount each weight is changed depends on its associated input value; that is, the blame for the error is meted out depending on which inputs had the most impact. For example, in the 8 of figure 3A, the higher-intensity (here, black) pixels would have the most impact, and the pixels with 0 intensity (here, white) would have no impact. (For interested readers, I have included some mathematical details in the notes.[17])

The whole process is repeated for the next training example. The training process goes through all the training examples multiple times, modifying the weights and threshold a little bit each time the perceptron makes an error. Just as the psychologist B. F. Skinner found when training pigeons, it's better to learn gradually over many trials; if the weights and threshold are changed too much on any one trial, then the system might end up learning the wrong thing (such as an overgeneralization that "the bottom and top halves of an 8 are always equal in size"). After many repetitions on each training example, the system eventually (we hope) settles on a set of weights and a threshold that result in correct answers for all the training examples. At that point, we can evaluate the perceptron on the test examples to see how it performs on images it hasn't been trained on.

An 8 detector is useful if you care only about 8s. But what about recognizing other digits? It's fairly straightforward to extend our perceptron to have ten outputs, one for each digit. Given an example handwritten digit, the output corresponding to that digit should be 1, and all the other outputs should be 0. This extended perceptron can learn all of its weights and thresholds using the perceptron-learning algorithm; the system just needs enough examples.

Rosenblatt and others showed that networks of perceptrons could learn to perform relatively simple perceptual tasks; moreover, Rosenblatt proved mathematically that for a certain, albeit very limited, class of tasks, perceptrons with sufficient training could, in principle, learn to perform these tasks without error. What wasn't clear was how well perceptrons could perform on more general AI tasks. This uncertainty didn't seem to stop Rosenblatt and his funders at the Office of Naval Research from making ridiculously optimistic predictions about their algorithm. Reporting on a press conference Rosenblatt held in July 1958, *The New York Times* featured this recap:

> The Navy revealed the embryo of an electronic computer today that it expects will be able to walk, talk, see, write, reproduce itself, and be conscious of its existence. Later perceptrons will be able to recognize people and call out their names and instantly translate speech in one language to speech and writing in another language, it was predicted.[18]

Yes, even at its beginning, AI suffered from a hype problem. I'll talk more about the unhappy results of such hype shortly. But for now, I want to use perceptrons to highlight a major difference between symbolic and sub-symbolic approaches to AI.

The fact that a perceptron's "knowledge" consists of a set of numbers—namely, the weights and threshold it has learned—means that it is hard to uncover the rules the perceptron is using in performing its recognition task. The perceptron's rules are not symbolic; unlike the General Problem Solver's symbols, such as LEFT-BANK, #MISSIONARIES, and MOVE, a perceptron's weights and threshold don't stand for particular concepts. It's not easy to translate these numbers into rules that are understandable by humans. The situation gets much worse with modern neural networks that have millions of weights.

One might make a rough analogy between perceptrons and the human brain. If I could open up your head and watch some subset of your hundred billion neurons firing, I would likely not get any insight into what you were thinking or the "rules" you used to make a particular decision. However, the human brain has given rise to language, which allows you to use symbols (words and phrases) to tell me—often imperfectly—what your thoughts are about or why you did a certain thing. In this sense, our neural firings can be considered *subsymbolic*, in that they underlie the symbols our brains somehow create. Perceptrons, as well as more complicated networks of simulated neurons, have been dubbed "subsymbolic" in analogy to the brain. Their advocates believe that to achieve artificial intelligence, language-like symbols and the rules that govern symbol processing cannot be programmed directly, as was done in the General Problem Solver, but must emerge from neural-like architectures similar to the way that intelligent symbol processing emerges from the brain.

The Limitations of Perceptrons

After the 1956 Dartmouth meeting, the symbolic camp dominated the AI landscape. In the early 1960s, while Rosenblatt was working avidly on the perceptron, the big four "founders" of AI, all strong devotees of the symbolic camp, had created influential—and well-funded—AI laboratories: Marvin Minsky at MIT, John McCarthy at Stanford, and Herbert Simon and Allen Newell at Carnegie Mellon. (Remarkably, these three universities remain to this day among the most prestigious places to study AI.) Minsky, in particular, felt that Rosenblatt's brain-inspired approach to AI was a dead end, and moreover was stealing away research dollars from more worthy symbolic AI efforts.[19] In 1969, Minsky and his MIT colleague Seymour Papert published a book, *Perceptrons*,[20] in which they gave a mathematical proof showing that the types of problems a perceptron could solve *perfectly* were very limited and that the perceptron-learning algorithm would not do well in scaling up to tasks requiring a large number of weights and thresholds.

Minsky and Papert pointed out that if a perceptron is augmented by adding a "layer" of simulated neurons, the types of problems that the device can solve is, in principle, much broader.[21] A perceptron with such an added layer is called a multilayer neural network. Such networks form the foundations of much of modern AI; I'll describe them in detail in the next chapter. But for now, I'll note that at the time of Minsky and Papert's book, multilayer neural networks were not broadly studied, largely because there was no general algorithm, analogous to the perceptron-learning algorithm, for learning weights and thresholds.

The limitations Minsky and Papert proved for simple perceptrons were already known to people working in this area.[22] Frank Rosenblatt himself had done extensive work on multilayer perceptrons and recognized the difficulty of training them.[23] It wasn't Minsky and Papert's mathematics that put the final nail in the perceptron's coffin; rather, it was their speculation on multilayer neural networks:

> [The perceptron] has many features to attract attention: its linearity; its intriguing learning theorem; its clear paradigmatic simplicity as a kind of parallel computation. There is no reason to suppose that any of these virtues carry over to the many-layered version. Nevertheless, we consider it to be an important research problem to elucidate (or reject) our intuitive judgment that the extension is sterile.[24]

Ouch. In today's vernacular that final sentence might be termed "passive-aggressive." Such negative speculations were at least part of the reason that funding for neural network research dried up in the late 1960s, at the same time that symbolic AI was flush with government dollars. In 1971, at the age of forty-three, Frank Rosenblatt died in a boating accident. Without its most prominent proponent, and without much government funding, research on perceptrons and other subsymbolic AI methods largely halted, except in a few isolated academic groups.

AI Winter

In the meantime, proponents of symbolic AI were writing grant proposals promising impending breakthroughs in areas such as speech and language understanding, commonsense reasoning, robot navigation, and autonomous vehicles. By the mid-1970s, while some very narrowly focused expert systems were successfully deployed, the more general AI breakthroughs that had been promised had not materialized.

The funding agencies noticed. Two reports, solicited respectively by the Science Research Council in the U.K. and the Department of Defense in the United States, reported very negatively on the progress and prospects for AI research. The U.K. report in particular acknowledged that there was promise in the area of specialized expert systems—"programs written to perform in highly specialised problem domains, when the programming takes very full account of the results of human experience and human intelligence within the relevant domain"—but concluded that the results to date were "wholly discouraging about general-purpose programs seeking to mimic the problem-solving aspects of human [brain] activity over a rather wide field. Such a general-purpose program, the coveted long-term goal of AI activity, seems as remote as ever."[25] This report led to a sharp decrease in government funding for AI research in the U.K.; similarly, the Department of Defense drastically cut funding for basic AI research in the United States.

This was an early example of a repeating cycle of bubbles and crashes in the field of AI. The two-part cycle goes like this. Phase 1: New ideas create a lot of optimism in the research community. Results of imminent AI breakthroughs are promised, and often hyped in the news media. Money pours in from government funders and venture capitalists for both academic research and commercial start-ups. Phase 2: The promised breakthroughs don't occur, or are much less impressive than promised. Government funding and venture capital dry up. Start-up companies fold, and AI research slows. This pattern became familiar to the AI community: "AI spring," followed by overpromising and media hype, followed by "AI winter." This has happened, to various degrees, in cycles of five to ten years. When I got out of graduate school in 1990, the field was in one of its winters and had garnered such a bad image that I was even advised to leave the term "artificial intelligence" off my job applications.

Easy Things Are Hard

The cold AI winters taught practitioners some important lessons. The simplest lesson was noted by John McCarthy, fifty years after the Dartmouth conference: "AI was harder than we thought."[26] Marvin Minsky pointed out that in fact AI research had uncovered a paradox: "Easy things are hard." The original goals of AI—computers that could converse with us in natural language, describe what they saw through their camera eyes, learn new concepts after seeing only a few examples—are things that young children can easily do, but, surprisingly, these "easy things" have turned out to be harder for AI to achieve than diagnosing complex diseases, beating human champions at chess and Go, and solving complex algebraic problems. As Minsky went on, "In general, we're least aware of what our minds do best."[27] The attempt to create artificial intelligence has, at the very least, helped elucidate how complex and subtle are our own minds.

2

·

Neural Networks

·

and the Ascent

·

# of Machine Learning

·

Spoiler alert: Multilayer neural networks—the extension of perceptrons that was dismissed by Minsky and Papert as likely to be "sterile"—have instead turned out to form the foundation of much of modern artificial intelligence. Because they are the basis of several of the methods I'll describe in later chapters, I'll take some time here to describe how these networks work.

Multilayer Neural Networks

A *network* is simply a set of elements that are connected to one another in various ways. We're all familiar with social networks, in which the elements are people, and computer networks, in which the elements are, naturally,

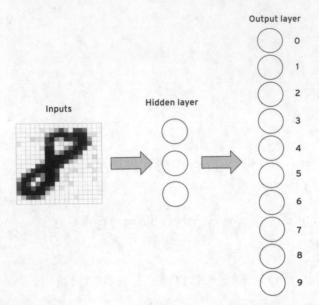

Output layer

0
1
2
3
4
5
6
7
8
9

Inputs

Hidden layer

FIGURE 4: **A two-layer neural network for recognizing handwritten digits**

computers. In neural networks, the elements are simulated neurons akin to the perceptrons I described in the previous chapter.

In figure 4, I've sketched a simple multilayer neural network, designed to recognize handwritten digits. The network has two columns (*layers*) of perceptron-like simulated neurons (circles). For simplicity (and probably to the relief of any neuroscientists reading this), I'll use the term *unit* instead of *simulated neuron* to describe the elements of this network. Like the 8-detecting perceptron from chapter 1, the network in figure 4 has 324 (18×18) inputs, each of which is set to the intensity value of the corresponding pixel in the input image. But unlike the perceptron, this network has a layer of three so-called hidden units, along with its layer of ten output units. Each output unit corresponds to one of the possible digit categories.

The large gray arrows signify that each input has a weighted connection to each hidden unit, and each hidden unit has a weighted connection to each output unit. The mysterious-sounding term *hidden unit* comes from the neural network literature; it simply means a non-output unit. A better name might have been *interior unit*.

Think of the structure of your brain, in which some neurons directly control "outputs" such as your muscle movements but most neurons simply communicate with other neurons. These could be called the brain's hidden neurons.

The network shown in figure 4 is referred to as "multilayered" because it has two layers of units (hidden and output) instead of just an output layer. In principle, a multilayer network can have multiple layers of hidden units; networks that have more than one layer of hidden units are called deep networks. The "depth" of a network is simply its number of hidden layers. I'll have much more to say about deep networks in upcoming chapters.

Similar to perceptrons, each unit here multiplies each of its inputs by the weight on that input's connection and then sums the results. However, unlike in a perceptron, a unit here doesn't simply "fire" or "not fire" (that is, produce 1 or 0) based on a threshold; instead, each unit uses its sum to compute a number between 0 and 1 that is called the unit's "activation." If the sum that a unit computes is low, the unit's activation is close to 0; if the sum is high, the activation is close to 1. (For interested readers, I've included some of the mathematical details in the notes.[1])

To process an image such as the handwritten 8 in figure 4, the network performs its computations layer by layer, from left to right. Each hidden unit computes its activation value; these activation values then become the inputs for the output units, which then compute their own activations. In the network of figure 4, the activation of an output unit can be thought of as the network's confidence that it is "seeing" the corresponding digit; the digit category with the highest confidence can be taken as the network's answer—its *classification*.

In principle, a multilayer neural network can learn to use its hidden units to recognize more abstract features (for example, visual shapes, such as the top and bottom "circles" on a handwritten 8) than the simple features (for example, pixels) encoded by the input. In general, it's hard to know ahead of time how many layers of hidden units are needed, or how many hidden units should be included in a layer, for a network to perform well on a given task. Most neural network researchers use a form of trial and error to find the best settings.

Learning via Back-Propagation

In their book *Perceptrons*, Minsky and Papert were skeptical that a successful algorithm could be designed for learning the weights in a multilayer neural network. Their skepticism (along with doubts from others in the symbolic AI community) was largely responsible for the sharp decrease in funding for neural network research in the 1970s. But despite the chilling effect of Minsky and Papert's book on the field, a small core of neural network researchers persisted, especially in Frank Rosenblatt's own field of cognitive psychology. And by the late 1970s and early '80s, several of these groups had definitively rebutted Minsky and Papert's speculations on the "sterility" of multilayer neural networks by developing a general learning algorithm—called back-propagation—for training these networks.

As its name implies, back-propagation is a way to take an error observed at the output units (for example, a high confidence for the wrong digit in the example of figure 4) and to "propagate" the blame for that error backward (in figure 4, this would be from right to left) so as to assign proper blame to each of the weights in the network. This allows back-propagation to determine how much to change each weight in order to reduce the error. *Learning* in neural networks simply consists in gradually modifying the weights on connections so that each output's error gets as close to 0 as possible on all training examples. While the mathematics of back-propagation is beyond the scope of my discussion here, I've included some details in the notes.[2]

Back-propagation will work (in principle at least) no matter how many inputs, hidden units, or output units your neural network has. While there is no mathematical guarantee that back-propagation will settle on the correct weights for a network, in practice it has worked very well on many tasks that are too hard for simple perceptrons. For example, I trained both a perceptron and a two-layer neural network, each with 324 inputs and 10 outputs, on the handwritten-digit-recognition task, using sixty thousand examples, and then tested how well each was able to recognize ten thousand new examples. The perceptron was correct on about 80 percent of the new examples, whereas the neural network, with 50 hidden units, was correct on a whopping 94 percent of those new examples. Kudos to the hidden units! But what exactly has the neural network learned that allowed it to soar past the perceptron? I don't know. It's possible that I could find a way to visualize the neural network's 16,700 weights[3] to get some insight into its performance, but I haven't done so, and in general it's not at all easy to understand how these networks make their decisions.

It's important to note that while I've used the example of handwritten digits, neural networks can be applied not just to images but to any kind of data. Neural networks have been applied in areas as diverse as speech recognition, stock-market prediction, language translation, and music composition.

Connectionism

In the 1980s, the most visible group working on neural networks was a team at the University of California at San Diego headed by two psychologists, David Rumelhart and James McClelland. What we now call neural networks were then generally referred to as connectionist networks, where the term *connectionist* refers to the idea that knowledge in these networks resides in weighted *connections* between units. The team led by Rumelhart and McClelland is known for writing the so-called bible of connectionism—a two-volume treatise, published in 1986, called *Parallel Distributed Processing*. In the midst of an AI landscape dominated by symbolic AI, the book was a pep talk for the subsymbolic approach, arguing that "people are smarter than today's computers because the brain employs a basic computational architecture that is more suited to . . . the natural information-processing tasks that people are so good at," for example, "perceiving objects in natural scenes and noting their relations, . . . understanding language, and retrieving contextually appropriate information from memory."[4] The authors speculated that "symbolic systems such as those favored by Minsky and Papert"[5] would not be able to capture these humanlike abilities.

Indeed, by the mid-1980s, expert systems—symbolic AI approaches that rely on humans to create rules that reflect expert knowledge of a particular domain—were increasingly revealing themselves to be *brittle*: that is, error-prone and often unable to generalize or adapt when presented with new situations. In analyzing the limitations of these systems, researchers were discovering how much the human experts writing the rules actually rely on subconscious knowledge—what you might call common sense—in order to act intelligently. This kind of common sense could not easily be captured in programmed rules or logical deduction, and the lack of it severely limited any broad application of symbolic AI methods. In short, after a cycle of grand promises, immense funding, and media hype, symbolic AI was facing yet another AI winter.

According to the proponents of connectionism, the key to intelligence was an appropriate computational architecture—inspired by the brain—and the ability of the system to learn on its own from data or from acting in the world. Rumelhart, McClelland, and their team constructed connectionist networks (in software) as scientific models of human learning, perception, and language development. While these networks did not exhibit anywhere near human-level performance, the various networks described in the *Parallel Distributed Processing* books and elsewhere were interesting enough as AI artifacts that many people took notice, including those at funding agencies. In 1988, a top official at the Defense Advanced Research Projects Agency (DARPA), which provided the lion's share of AI funding, proclaimed, "I believe that this technology which we are about to embark upon [that is, neural networks] is more important than the atom bomb."[6] Suddenly neural networks were "in" again.

Bad at Logic, Good at Frisbee

Over the last six decades of AI research, people have repeatedly debated the relative advantages and disadvantages of symbolic and subsymbolic approaches. Symbolic systems can be engineered by humans, be imbued with human knowledge, and use human-understandable reasoning to solve problems. For example, MYCIN, an expert system developed in the early 1970s, was given about six hundred rules that it used to help physicians diagnose and treat blood diseases. MYCIN's programmers developed these rules after painstaking interviews with expert physicians. Given a patient's symptoms and medical test results, MYCIN was able to use both logic and probabilistic reasoning together with its rules in order to come up with a diagnosis, and it was able to explain its reasoning process. In short, MYCIN was a paradigmatic example of symbolic AI.

In contrast, as we've seen, subsymbolic systems tend to be hard to interpret, and no one knows how to directly program complex human knowledge or logic into these systems. Subsymbolic systems seem much better suited to perceptual or motor tasks for which humans can't easily define rules. You can't easily write down rules for identifying handwritten digits, catching a baseball, or recognizing your mother's voice; you just seem to do it automatically, without conscious thought. As the philosopher Andy Clark put it, the nature of subsymbolic systems is to be "bad at logic, good at Frisbee."[7]

So, why not just use symbolic systems for tasks that require high-level language-like descriptions and logical reasoning, and use subsymbolic systems for the low-level perceptual tasks such as recognizing faces and voices? To some extent, this is what has been done in AI, with very little connection between the two areas. Each of these approaches has had important successes in narrow areas but has serious limitations in achieving the original goals of AI. While there have been some attempts to construct hybrid systems that integrate subsymbolic and symbolic methods, none have yet led to any striking success.

The Ascent of Machine Learning

Inspired by statistics and probability theory, AI researchers developed numerous algorithms that enable computers to learn from data, and the field of machine learning became its own independent subdiscipline of AI, intentionally separate from symbolic AI. Machine-learning researchers disparagingly referred to symbolic AI methods as good old-fashioned AI, or GOFAI (pronounced "go-fye"),[8] and roundly rejected them.

Over the next two decades, machine learning had its own cycles of optimism, government funding, start-ups, and overpromising, followed by the inevitable winters. Training neural networks and similar methods to solve real-world problems could be glacially slow, and often didn't work very well, given the limited amount of data and computer power available at the time. But more data and computing power were coming shortly. The explosive growth of the internet would see to that. The stage was set for the next big AI revolution.

3

.

AI Spring

.

Spring Fever

Have you ever taken a video of your cat and uploaded it to YouTube? If so, you are not alone. More than a billion videos have been uploaded to YouTube, and a lot of them feature cats. In 2012, an AI team at Google constructed a multilayer neural network with over a billion weights that "viewed" millions of random YouTube videos while it adjusted these weights in order to successfully compress, and then decompress, selected frames from the videos. The Google researchers didn't tell the system to learn about any particular objects, but after a week of training, when they probed the innards of the network, what did they find? A "neuron" (unit) that seemed to encode cats.[1] This self-taught cat-recognition machine was one of a series of impressive AI feats that have captured the public's attention over the last decade. Most of these achievements rely on a set of neural network algorithms known as deep learning.

Until recently, AI's popular image came largely from the many movies and TV shows in which it played a starring role; think *2001: A Space Odyssey* or *The Terminator*. Real-world AI wasn't very noticeable in our everyday lives or mainstream media. If you came of age in the 1990s or earlier, you might recall frustrating encounters with customer service speech-recognition systems, the robotic word-learning toy Furby, or Microsoft's annoying and ill-fated Clippy, the paper-clip virtual assistant. Full-blown AI didn't seem imminent.

Maybe this is why so many people were shocked and upset when, in 1997, IBM's Deep Blue chess-playing system defeated the world chess champion Garry Kasparov. This event so stunned Kasparov that he accused the IBM team of cheating; he assumed that for the machine to play so well, it must have received help from human experts.[2] (In a nice bit of irony, during the 2006 World Chess Championship matches the tables were turned, with one player accusing the other of cheating by receiving help from a computer chess program.[3])

Our collective human angst over Deep Blue quickly receded. We accepted that chess could yield to brute-force machinery; playing chess well, we allowed, didn't require general intelligence after all. This seems to be a common response when computers surpass humans on a particular task; we conclude that the task doesn't actually require intelligence. As John McCarthy lamented, "As soon as it works, no one calls it AI anymore."[4]

However, by the mid-2000s and beyond, a more pervasive succession of AI accomplishments started sneaking up on us and then proliferating at a dizzying pace. Google launched its automated language-translation service, Google Translate. It wasn't perfect, but it worked surprisingly well, and it has since improved significantly. Shortly thereafter, Google's self-driving cars showed up on the roads of Northern California, careful and timid, but commuting on their own in full traffic. Virtual assistants such as Apple's Siri and Amazon's Alexa were installed on our phones and in our homes and could deal with many of our spoken requests. YouTube started providing impressively accurate automated subtitles for videos, and Skype offered simultaneous translation between languages in video calls. Suddenly Facebook could recognize your face eerily well in uploaded photos, and the photo-sharing website Flickr began automatically labeling photos with text describing their content.

In 2011, IBM's Watson program roundly defeated human champions on television's *Jeopardy!* game show, adroitly interpreting pun-laden clues and prompting its challenger Ken Jennings to "welcome our new computer overlords." Just five years later, millions of internet viewers were introduced to the complex game of Go, a longtime grand challenge for AI, when a program called AlphaGo stunningly defeated one of the world's best players in four out of five games.

The buzz over artificial intelligence was quickly becoming deafening, and the commercial world took notice. All of the largest technology companies have poured billions of dollars into AI research and development, either hiring AI experts directly or acquiring smaller start-up companies for the sole purpose of grabbing ("acqui-hiring") their talented employees. The potential of being acquired, with its promise of instant millionaire status, has fueled a proliferation of start-ups, often founded and run by former university professors, each with his or her own twist on AI. As the technology journalist Kevin Kelly observed, "The business plans of the next 10,000 startups are easy to forecast: Take X and add AI."[5] And, crucially, for nearly all of these companies, *AI* has meant "deep learning."

AI spring is once again in full bloom.

AI: Narrow and General, Weak and Strong

Like every AI spring before it, our current one features experts predicting that "general AI"—AI that equals or surpasses humans in most ways—will be here soon. "Human level AI will be passed in the mid-2020s,"[6] predicted Shane Legg, cofounder of Google DeepMind, in 2008. In 2015, Facebook's CEO, Mark Zuckerberg, declared, "One of our goals for the next five to 10 years is to basically get better than human level at all of the primary human senses: vision, hearing, language, general cognition."[7] The AI philosophers Vincent Müller and Nick Bostrom published a 2013 poll of AI researchers in which many assigned a 50 percent chance of human-level AI by the year 2040.[8]

While much of this optimism is based on the recent successes of deep learning, these programs—like *all* instances of AI to date—are still examples of what is called "narrow" or "weak" AI. These terms are not as derogatory as they sound; they simply refer to a system that can perform only one narrowly defined task (or a small set of related tasks). AlphaGo is possibly the world's best Go player, but it can't do anything else; it can't even play checkers, tic-tac-toe, or Candy Land. Google Translate can render an English movie review into Chinese, but it can't tell you if the reviewer liked the movie or not, and it certainly can't watch and review the movie itself.

The terms *narrow* and *weak* are used to contrast with *strong, human-level, general,* or *full-blown* AI (sometimes called AGI, or artificial general intelligence)—that is, the AI that we see in movies, that can do most everything we humans can do, and possibly much more. General AI might have been the original goal of the field, but achieving it has turned out to be much harder than expected. Over time, efforts in AI have become focused on particular well-defined tasks—speech recognition, chess playing, autonomous driving, and so on. Creating machines that perform such functions is useful and often lucrative, and it could be argued that each of these tasks individually requires "intelligence." But no AI program has been created yet that could be called intelligent in any general sense. A recent appraisal of the field stated this well: "A pile of narrow intelligences will never add up to a general intelligence. General intelligence isn't about the number of abilities, but about the integration between those abilities."[9]

But wait. Given the rapidly increasing pile of narrow intelligences, how long will it be before someone figures out how to integrate them and produce all of the broad, deep, and subtle features of human intelligence? Do we believe the cognitive scientist Steven Pinker, who thinks all this is business as usual? "Human-level AI is still the standard fifteen to twenty-five years away, just as it always has been, and many of its recently touted advances have shallow roots," Pinker declared.[10] Or should we pay more attention to the AI optimists, who are certain that this time around, this AI spring, things will be different?

Not surprisingly, in the AI research community there is considerable controversy over what human-level AI would entail. How can we know if we have succeeded in building such a "thinking machine"? Would such a system be required to have *consciousness* or *self-awareness* in the way humans do? Would it need to *understand* things in the same way a human understands them? Given that we're talking about a machine here, would we be more correct to say it is "simulating thought," or could we say it is truly thinking?

Could Machines Think?

Such philosophical questions have dogged the field of AI since its inception. Alan Turing, the British mathematician who in the 1930s sketched out the first framework for programmable computers, published a paper in 1950 asking what we might mean when we ask, "Can machines think?" After proposing his famous "imitation game" (now called the Turing test—more on this in a bit), Turing listed nine possible objections to the prospect of a machine actually thinking, all of which he tried to refute. These imagined objections range from the theological—"Thinking is a function of man's immortal soul. God has given an immortal soul to every man and woman, but not to any other animal or to machines. Hence no animal or machine can think"—to the parapsychological, something along the lines of "Humans can use telepathy to communicate while machines cannot." Strangely enough, Turing judged this last argument as "quite a strong one," because "the statistical evidence, at least for telepathy, is overwhelming."

From the vantage of many decades, my own vote for the strongest of Turing's possible arguments is the "argument from consciousness," which he summarizes by quoting the neurologist Geoffrey Jefferson:

> Not until a machine can write a sonnet or compose a concerto because of thoughts and emotions felt, and not by the chance fall of symbols, could we agree that machine equals brain—that is, not only write it but know

that it had written it. No mechanism could feel (and not merely artificially signal, an easy contrivance) pleasure at its successes, grief when its valves fuse, be warmed by flattery, be made miserable by its mistakes, be charmed by sex, be angry or depressed when it cannot get what it wants.[11]

Note that this argument is saying the following: (1) Only when a machine *feels* things and is aware of its own actions and feelings—in short, is conscious—could we consider it actually *thinking*, and (2) No machine could ever do this. Ergo, no machine could ever *actually* think.

I think it's a strong argument, even though I don't agree with it. It resonates with our intuitions about what machines are and how they are limited. Over the years, I've talked with any number of friends, relatives, and students about the possibility of machine intelligence, and this is the argument many of them stand by. For example, I was recently talking with my mother, a retired lawyer, after she had read a *New York Times* article about advances in the Google Translate program:

MOM: *The problem with people in the field of AI is that they anthropomorphize so much!*

ME: *What do you mean, anthropomorphize?*

MOM: *The language they use implies that machines might be able to actually think, rather than to just simulate thinking.*

ME: *What's the difference between "actually thinking" and "simulating thinking"?*

MOM: *Actual thinking is done with a brain, and simulating is done with computers.*

ME: *What's so special about a brain that it allows "actual" thinking? What's missing in computers?*

MOM: *I don't know. I think there's a human quality to thinking that can't ever be completely mimicked by computers.*

My mother isn't the only one who has this intuition. In fact, to many people it seems so obvious as to require no argument. And like many of these people, my mother would claim to be a philosophical materialist; that is, she doesn't believe in any nonphysical "soul" or "life force" that imbues living things with intelligence. It's just that she doesn't think machines could ever have the right stuff to "actually think."

In the academic realm, the most famous version of this argument was put forth by the philosopher John Searle. In 1980, Searle published an article called "Minds, Brains, and Programs"[12] in which he vigorously argued against the possibility of machines *actually thinking*. In this widely read, controversial piece, Searle introduced the concepts of "strong" and "weak" AI in order to distinguish between two philosophical claims made about AI programs. While many people today use the phrase *strong AI* to mean "AI that can perform most tasks as well as a human" and *weak AI* to mean the kind of narrow AI that currently exists, Searle meant something different by these terms. For Searle, the *strong AI* claim would be that "the appropriately programmed digital computer does not just simulate having a mind; it literally has a mind."[13] In contrast, in Searle's terminology, *weak AI* views computers as tools to simulate human intelligence and does not make any claims about them "literally" having a mind.[14] We're back to the philosophical question I was discussing with my mother: Is there a difference between "simulating a mind" and "literally having a mind"? Like my mother, Searle believes there is a fundamental difference, and he argued that strong AI is impossible even in principle.[15]

The Turing Test

Searle's article was spurred in part by Alan Turing's 1950 paper, "Computing Machinery and Intelligence," which had proposed a way to cut through the Gordian knot of "simulated" versus "actual" intelligence. Declaring that "the original question 'Can a machine think?' is too meaningless to deserve discussion," Turing proposed an operational method to give it meaning. In his "imitation game," now called the Turing test, there are two contestants: a computer and a human. Each is questioned separately by a (human) judge who tries to determine which is which. The judge is physically separated from the two contestants so cannot rely on visual or auditory cues; only typed text is communicated.

Turing suggested the following: "The question, 'Can machines think?' should be replaced by 'Are there imaginable digital computers which would do well in the imitation game?'" In other words, if a computer is sufficiently humanlike to be indistinguishable from humans, aside from its physical appearance or what it sounds like (or smells or feels like, for that matter), why shouldn't we consider it to *actually* think? Why should we require an entity to be created out of a particular kind of material (for example, biological cells) to grant it "thinking" status? As the computer scientist Scott Aaronson put it bluntly, Turing's proposal is "a plea against meat chauvinism."[16]

The devil is always in the details, and the Turing test is no exception. Turing did not specify the criteria for selecting the human contestant and the judge, or stipulate how long the test should last, or what conversational topics should be allowed. However, he did make an oddly specific prediction: "I believe that in about 50 years' time it will be possible to programme computers . . . to make them play the imitation game so well that an average interrogator will not have more than 70 percent chance of making the right identification after five minutes of questioning." In other words, in a five-minute session, the average judge will be fooled 30 percent of the time.

Turing's prediction has turned out to be pretty accurate. Several Turing tests have been staged over the years, in which the computer contestants are chatbots—programs specifically built to carry on conversations (they can't do anything else). In 2014, the Royal Society in London was host to a Turing test demonstration featuring five computer programs, thirty human contestants, and thirty human judges of different ages and walks of life, including computer experts and nonexperts, as well as native and nonnative English speakers. Each judge conducted several rounds of five-minute conversations in which he or she conversed (by typing) in parallel with a pair of contestants—one human and one machine—after which the judge had to guess which was which. A chatbot named "Eugene Goostman," created by a group of Russian and Ukrainian programmers, won the competition by fooling ten (or 33.3 percent) of the judges. The competition organizers, following Turing's "more than 30 percent fooled in five minutes" criterion, quickly flooded the media with reports that the Turing test had finally been passed.

The Eugene Goostman chatbot works in much the same way as most other chatbots. It stores a large set of sentence templates that can be filled in based on a set of programmed rules that are applied to the input text it gets from its conversation partner. The chatbot's programmers have given it linguistic rules that allow it to pinpoint key information in its input and to store that information for later use. In addition, the chatbot stores a database of "commonsense knowledge," encoded by human programmers, along with some logic rules; these would presumably allow it to answer questions like "What color is the sky over Vladivostok?" with the human-sounding "Blue, I suppose, unless it's a cloudy day." If none of the chatbot's rules apply to an input, it just changes the subject. The system's rules also encode its "personality"—in this case, a thirteen-year-old Ukrainian boy whose English is good but (conveniently) not perfect. Here is a sample of Eugene Goostman's conversation with one of the judges who was fooled:

> JUDGE: *What is your gender*
> EUGENE: *I'm a young boy, if you care to know. And not very ugly, by the way!*
> JUDGE: *what are you aware of at this moment*
> EUGENE: *I would rather not talk about it if you don't mind. Could you tell me what are you? I mean your profession.*
> JUDGE: *how old are you*
> EUGENE: *I'm a little boy 13 years old.*
> JUDGE: *what are your qualifications*
> EUGENE: *My qualifications? I'm not gonna tell you. Oooh. Anything else?*[17]

After the competition, the organizers issued a press release announcing, "The 65-year-old iconic Turing Test was passed for the very first time by computer programme Eugene Goostman," and stating, "It is fitting that such an important landmark has been reached at the Royal Society in London, the home of British Science and the scene of many great advances in human understanding over the centuries. This milestone will go down in history as one of the most exciting."[18]

AI experts unanimously scoffed at this characterization. To anyone familiar with how chatbots are programmed, it's glaringly obvious from the competition transcripts that Eugene Goostman is a program, and not even a very sophisticated one. The result seemed to reveal more about the judges and the test itself than about the machines. Given five minutes and a propensity to avoid hard questions by changing the subject or by responding with a new question, the program had a surprisingly easy time fooling a nonexpert judge into believing he or she was conversing with a real person. This has been demonstrated with many chatbots, ranging from the 1960s ELIZA, which mimicked a psychotherapist, to today's malevolent Facebook bots, which use short text exchanges to trick people into revealing personal information.

These bots are, of course, leveraging our very human tendency to anthropomorphize (you were right, Mom!). We are all too willing to ascribe understanding and consciousness to computers, based on little evidence.

For these reasons, most AI experts hate the Turing test, at least as it has been carried out to date. They see such competitions as publicity stunts whose results say nothing about progress in AI. But while Turing might have overestimated the ability of an "average interrogator" to see through superficial trickery, could the test still be a useful indicator of actual intelligence if the conversation time is extended and the required expertise of the judges is raised?

Ray Kurzweil, who is now director of engineering at Google, believes that a properly designed version of the Turing test will indeed reveal machine intelligence; he predicts that a computer will pass this test by 2029, a milestone event on the way to Kurzweil's forecasted Singularity.

The Singularity

Ray Kurzweil has long been AI's leading optimist. A former student of Marvin Minsky's at MIT, Kurzweil has had a distinguished career as an inventor: he invented the first text-to-speech machine as well as one of the world's best music synthesizers. In 1999, President Bill Clinton awarded Kurzweil the National Medal of Technology and Innovation for these and other inventions.

Yet Kurzweil is best known not for his inventions but for his futurist prognostications, most notably the idea of the Singularity: "a future period during which the pace of technological change will be so rapid, its impact so deep, that human life will be irreversibly transformed."[19] Kurzweil uses the term *singularity* in the sense of "a unique event with . . . singular implications"; in particular, "an event capable of rupturing the fabric of human history."[20] For Kurzweil, this singular event is the point in time when AI exceeds human intelligence.

Kurzweil's ideas were spurred by the mathematician I. J. Good's speculations on the potential of an intelligence explosion: "Let an ultraintelligent machine be defined as a machine that can far surpass all the intellectual activities of any man however clever. Since the design of machines is one of these intellectual activities, an ultraintelligent machine could design even better machines; there would then unquestionably be an 'intelligence explosion,' and the intelligence of man would be left far behind."[21]

Kurzweil was also influenced by the mathematician and science fiction writer Vernor Vinge, who believed this event was close at hand: "The evolution of human intelligence took millions of years. We will devise an equivalent advance in a fraction of that time. We will soon create intelligences greater than our own. When this happens, human history will have reached a kind of singularity . . . and the world will pass far beyond our understanding."[22]

Kurzweil takes the intelligence explosion as his starting point and then turns up the sci-fi intensity, moving from AI to nanoscience, then to virtual reality and "brain uploading," all in the same calm, confident tone of a Delphic oracle looking at a calendar and pointing to specific dates. To give you the flavor of all this, here are some of Kurzweil's predictions:

> By the 2020s molecular assembly will provide tools to effectively combat poverty, clean up our environment, overcome disease, [and] extend human longevity.

> By the end of the 2030s . . . brain implants based on massively distributed intelligent nanobots will greatly expand our memories and otherwise vastly improve all our sensory, pattern-recognition, and cognitive abilities.

> Uploading a human brain means scanning all of its salient details and then reinstantiating those details into a suitably powerful computational

substrate. . . . The end of the 2030s is a conservative projection for successful [brain] uploading.[23]

A computer will pass the Turing test by 2029.[24]

As we get to the 2030s, artificial consciousness will be very realistic.
That's what it means to pass the Turing test.[25]

I set the date for the Singularity . . . as 2045. The nonbiological intelligence created in that year will be one billion times more powerful than
all human intelligence today.[26]

The writer Andrian Kreye wryly referred to Kurzweil's Singularity prediction as "nothing more than the belief in a technological Rapture."[27]

Kurzweil bases all of his predictions on the idea of "exponential progress" in many areas of science and technology, especially computers. To
unpack this idea, let's consider how exponential growth works.

An Exponential Fable

For a simple illustration of exponential growth, I'll recount an old fable.
Long ago, a renowned sage from a poor and starving village visited a distant and rich kingdom where the king challenged him to a game of chess.
The sage was reluctant to accept, but the king insisted, offering the sage
a reward "of anything you desire, if you are able to defeat me in a game."
For the sake of his village, the sage finally accepted and (as sages usually
do) won the game. The king asked the sage to name his reward. The sage,
who enjoyed mathematics, said, "All I ask for is that you take this chessboard, put two grains of rice on the first square, four grains on the second
square, eight grains on the third, and so on, doubling the number of grains
on each successive square. After you complete each row, package up the
rice on that row and ship it to my village." The mathematically naive king
laughed. "Is that all you want? I will have my men bring in some rice and
fulfill your request posthaste."

The king's men brought in a large bag of rice. After several minutes they had completed the first eight squares of the board with the requisite grains of rice: 2 on the first square, 4 on the second, 8 on the third, and so on, with 256 grains on the eighth square. They put the collection of grains (511, to be exact) in a tiny bag and sent it off by horseback to the sage's village. They then proceeded on to the second row, with 512 grains on the first square of that row, 1,024 grains on the next square, and 2,048 grains on the following. Each pile of rice no longer fit on a chessboard square, so it was counted into a large bowl instead. By the end of the second row, the counting of grains was taking far too much time, so the court mathematicians started estimating the amounts by weight. They calculated that for the sixteenth square, 65,536 grains—about a kilogram (just over two pounds)—were required. The bag of rice shipped off for the second row weighed about two kilograms.

The king's men started on the third row. The seventeenth square required 2 kilos, the eighteenth required 4, and so on; by the end of the third row (square 24), 512 kilos were needed. The king's subjects were conscripted to bring in additional giant bags of rice. The situation had become dire by the second square of the fourth row (square 26), when the mathematicians calculated that 2,048 kilos (over two tons) of rice were required. This would exhaust the entire rice harvest of the kingdom, even though the chessboard was not even half completed. The king, now realizing the trick that had been played on him, begged the sage to relent and save the kingdom from starvation. The sage, satisfied that the rice already received by his village would be enough, agreed.

Figure 5A plots the number of kilos of rice required on each chess square, up to the twenty-fourth square. The first square, with two rice grains, has a scant fraction of a kilo. Similarly, the squares up through 16 have less than 1 kilo. But after square 16, you can see the plot shoot up rapidly, due to the doubling effect. Figure 5B shows the values for the twenty-fourth through the sixty-fourth chess square, going from 512 kilos to more than 30 trillion kilos.

The mathematical function describing this graph is $y = 2^x$, where x is the chess square (numbered from 1 to 64) and y is the number of rice grains required on that square. This is called an exponential function, because x is the exponent of the number 2. No matter what scale is plotted,

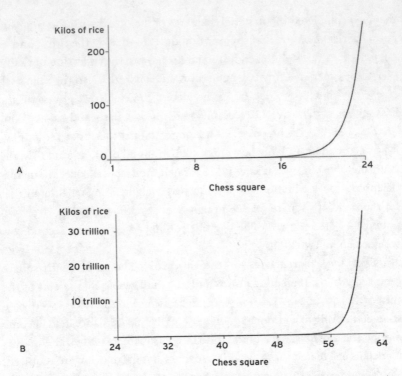

FIGURE 5: Plots showing how many kilos of rice are needed for each chess square in order to fulfill the sage's request; A, squares 1–24 (with y-axis showing hundreds of kilos); B, squares 24–64 (with y-axis showing tens of trillions of kilos)

the function will have a characteristic point at which the curve seems to change from slow to explosively fast growth.

Exponential Progress in Computers

For Ray Kurzweil, the computer age has provided a real-world counterpart to the exponential fable. In 1965, Gordon Moore, cofounder of Intel Corporation, identified a trend that has come to be known as Moore's law: the number of components on a computer chip doubles approximately every one to two years. In other words, the components are getting exponentially smaller (and cheaper), and computer speed and memory are increasing at an exponential rate.

Kurzweil's books are full of graphs like the ones in figure 5, and extrapolations of these trends of exponential progress, along the lines of Moore's law, are at the heart of his forecasts for AI. Kurzweil points out that if the trends continue (as he believes they will), a $1,000 computer will "achieve human brain capability (10^{16} calculations per second) . . . around the year 2023."[28] At that point, in Kurzweil's view, human-level AI will just be a matter of reverse engineering the brain.

Neural Engineering

Reverse engineering the brain means understanding enough about its workings in order to duplicate it, or at least to use the brain's underlying principles to replicate its intelligence in a computer. Kurzweil believes that such reverse engineering is a practical, near-term approach to creating human-level AI. Most neuroscientists would vehemently disagree, given how little is currently known about how the brain works. But Kurzweil's argument again rests on exponential trends—this time in advancements in neuroscience. In 2002 he wrote, "A careful analysis of the requisite trends shows that we will understand the principles of operation of the human brain and be in a position to recreate its powers in synthetic substances well within thirty years."[29]

Few if any neuroscientists agree on this optimistic prediction for their field. But even if a machine operating on the brain's principles can be created, how will it learn all the stuff it needs to know to be considered intelligent? After all, a newborn baby has a brain, but it doesn't yet have what we'd call human-level intelligence. Kurzweil agrees: "Most of [the brain's] complexity comes from its own interaction with a complex world. Thus, it will be necessary to provide an artificial intelligence with an education just as we do with a natural intelligence."[30]

Of course, providing an education can take many years. Kurzweil thinks that the process can be vastly sped up. "Contemporary electronics is already more than ten million times faster than the human nervous system's electrochemical information processing. Once an AI masters human basic language skills, it will be in a position to expand its language skills and general knowledge by rapidly reading all human literature and by absorbing the knowledge contained on millions of web sites."[31]

Kurzweil is vague on how all this will happen but assures us that to achieve human-level AI, "we will not program human intelligence link by link as in some massive expert system. Rather, we will set up an intricate hierarchy of self-organizing systems, based largely on the reverse engineering of the human brain, and then provide for its education . . . hundreds if not thousands of times faster than the comparable process for humans."[32]

Singularity Skeptics and Adherents

Responses to Kurzweil's books *The Age of Spiritual Machines* (1999) and *The Singularity Is Near* (2005) are often one of two extremes: enthusiastic embrace or dismissive skepticism. When I read Kurzweil's books, I was (and still am) in the latter camp. I wasn't at all convinced by his surfeit of exponential curves or his arguments for reverse engineering the brain. Yes, Deep Blue had defeated Kasparov in chess, but AI was far below the level of humans in most other domains. Kurzweil's predictions that AI would equal us in a mere couple of decades seemed to me ridiculously optimistic.

Most of the people I know are similarly skeptical. Mainstream AI's attitude is perfectly captured in an article by the journalist Maureen Dowd: she describes how Andrew Ng, a famous AI researcher from Stanford, rolled his eyes at her mention of Kurzweil, saying, "Whenever I read Kurzweil's *Singularity*, my eyes just naturally do that."[33]

On the other hand, Kurzweil's ideas have many adherents. Most of his books have been bestsellers and have been positively reviewed in serious publications. *Time* magazine declared of the Singularity, "It's not a fringe idea; it's a serious hypothesis about the future of life on Earth."[34]

Kurzweil's thinking has been particularly influential in the tech industry, where people often believe in exponential technological progress as the means to solve all of society's problems. Kurzweil is not only a director of engineering at Google but also a cofounder (with his fellow futurist entrepreneur Peter Diamandis) of Singularity University (SU), a "trans-humanist" think tank, start-up incubator, and sometime summer camp for the tech elite. SU's published mission is "to educate, inspire, and empower leaders to apply exponential technologies to address humanity's grand challenges."[35] The organization is partially underwritten by Google;

Larry Page (cofounder of Google) was an early supporter and is a frequent speaker at SU's programs. Several other big-name technology companies have joined as sponsors.

Douglas Hofstadter is one thinker who—again surprising me—straddles the fence between Singularity skepticism and worry. He was disturbed, he told me, that Kurzweil's books "mixed in the zaniest science fiction scenarios with things that were very clearly true." When I argued, Hofstadter pointed out that from the vantage of several years later, for every seemingly crazy prediction Kurzweil made, he also often predicted something that has surprisingly come true or will soon. By the 2030s, will "'experience beamers' . . . send the entire flow of their sensory experiences as well as the neurological correlates of their emotional reactions out onto the Web"?[36] Sounds crazy. But in the late 1980s, Kurzweil, relying on his exponential curves, predicted that by 1998 "a computer will defeat the human world chess champion . . . and we'll think less of chess as a result."[37] At the time, many thought that sounded crazy too. But this event occurred a year earlier than Kurzweil predicted.

Hofstadter has noted Kurzweil's clever use of what Hofstadter calls the "Christopher Columbus ploy,"[38] referring to the Ira Gershwin song "They All Laughed," which includes the line "They all laughed at Christopher Columbus." Kurzweil cites numerous quotations from prominent people in history who completely underestimated the progress and impact of technology. Here are a few examples. IBM's chairman, Thomas J. Watson, in 1943: "I think there is a world market for maybe five computers." Digital Equipment Corporation's cofounder Ken Olsen in 1977: "There's no reason for individuals to have a computer in their home." Bill Gates in 1981: "640,000 bytes of memory ought to be enough for anybody."[39] Hofstadter, having been stung by his own wrong predictions on computer chess, was hesitant to dismiss Kurzweil's ideas out of hand, as crazy as they sounded. "Like Deep Blue's defeat of Kasparov, it certainly gives one pause for thought."[40]

Wagering on the Turing Test

As a career choice, "futurist" is nice work if you can get it. You write books making predictions that can't be evaluated for decades and whose ultimate validity won't affect your reputation—or your book sales—in the here and now. In 2002, a website called Long Bets was created to help keep futurists honest. Long Bets is "an arena for competitive, accountable predictions,"[41] allowing a *predictor* to make a long-term prediction that specifies a date and a *challenger* to challenge the prediction, both putting money on a wager that will be paid off after the prediction's date is passed. The site's very first predictor was the software entrepreneur Mitchell Kapor. He made a negative prediction: "By 2029 no computer—or 'machine intelligence'— will have passed the Turing Test." Kapor, who had founded the successful software company Lotus and who is also a longtime activist on internet civil liberties, knew Kurzweil well and was on the "highly skeptical" side of the Singularity divide. Kurzweil agreed to be the challenger for this public bet, with $20,000 going to the Electronic Frontier Foundation (cofounded by Kapor) if Kapor wins and to the Kurzweil Foundation if Kurzweil wins. The test to determine the winner will be carried out before the end of 2029.

In making this wager, Kapor and Kurzweil had to—unlike Turing— specify carefully in writing how their Turing test would work. They begin with a few necessary definitions. "A Human is a biological human person as that term is understood in the year 2001 whose intelligence has not been enhanced through the use of machine (i.e., nonbiological) intelligence. . . . A Computer is any form of nonbiological intelligence (hardware and software) and may include any form of technology, but may not be a biological Human (enhanced or otherwise) nor biological neurons (however, nonbiological emulations of biological neurons are allowed)."[42]

The terms of the wager also specify that the test will be carried out by three human judges who will interview the computer contestant as well as three human "foils." All four contestants will try to convince the judges that they are humans. The judges and human foils will be chosen by a "Turing test committee," made up of Kapor, Kurzweil (or their designees), and a third member. Instead of five-minute chats, each of the four contestants will be interviewed by each judge for a grueling two hours. At the end of all these interviews, each judge will give his or her verdict ("human" or "machine") for each contestant. "The Computer will be deemed to have passed the 'Turing Test Human Determination Test' if the

Computer has fooled two or more of the three Human Judges into think-ing that it is a human."[43]

But we're not done yet:

> In addition, each of the three Turing Test Judges will rank the four Can-didates with a rank from 1 (least human) to 4 (most human). The com-puter will be deemed to have passed the "Turing Test Rank Order Test" if the median rank of the Computer is equal to or greater than the me-dian rank of two or more of the three Turing Test Human Foils.

> The Computer will be deemed to have passed the Turing Test if it passes both the Turing Test Human Determination Test and the Turing Test Rank Order Test.

> If a Computer passes the Turing Test, as described above, prior to the end of the year 2029, then Ray Kurzweil wins the wager. Otherwise Mitchell Kapor wins the wager.[44]

Wow, pretty strict. Eugene Goostman wouldn't stand a chance. I'd have to (cautiously) agree with this assessment from Kurzweil: "In my view, there is no set of tricks or simpler algorithms (i.e., methods simpler than those underlying human intelligence) that would enable a machine to pass a properly designed Turing Test without actually possessing intelligence at a fully human level."[45]

In addition to laying out the rules of their long bet, both Kapor and Kurzweil wrote accompanying essays giving the reasons each thinks he will win. Kurzweil's essay summarizes the arguments laid out in his books: exponential progress in computation, neuroscience, and nanotech-nology, which taken together will allow for reverse engineering of the brain.

Kapor doesn't buy it. His main argument centers on the influence of our (human) physical bodies and emotions on our cognition. "Perception of and [physical] interaction with the environment is the equal partner of cognition in shaping experience. . . . [Emotions] bound and shape the en-velope of what is thinkable."[46] Kapor asserts that without the equivalent of a human body, and everything that goes along with it, a machine will never be able to learn all that's needed to pass his and Kurzweil's strict Turing test.

> I assert that the fundamental mode of learning of human beings is experiential. Book learning is a layer on top of that. . . . If human knowledge, especially knowledge about experience, is largely tacit, i.e., never directly and explicitly expressed, it will not be found in books, and the Kurzweil approach to knowledge acquisition will fail. . . . It is not in what the computer knows but what the computer does not know and cannot know wherein the problem resides.[47]

Kurzweil responds that he agrees with Kapor on the role of experiential learning, tacit knowledge, and emotions but believes that before the 2030s virtual reality will be "totally realistic,"[48] enough to re-create the physical experiences needed to educate a developing artificial intelligence. (Welcome to the Matrix.) Moreover, this artificial intelligence will have a reverse-engineered artificial brain with emotion as a key component.

Are you, like Kapor, skeptical of Kurzweil's predictions? Kurzweil says it's because you don't understand exponentials. "Generally speaking, the core of a disagreement I'll have with a critic is, they'll say, Oh Kurzweil is underestimating the complexity of reverse-engineering the human brain or the complexity of biology. But I don't believe I'm underestimating the challenge. I think they're underestimating the power of exponential growth."[49]

Kurzweil's doubters point out a couple of holes in this argument. Indeed, computer hardware has seen exponential progress over the last five decades, but there are many reasons to believe this trend will not hold up in the future. (Kurzweil of course disputes this.) But more important, computer *software* has not shown the same exponential progress; it would be hard to argue that today's software is exponentially more sophisticated, or brain-like, than the software of fifty years ago, or that such a trend has ever existed. Kurzweil's claims about exponential trends in neuroscience and virtual reality are also widely disputed.

But as Singularitarians have pointed out, sometimes it's hard to see an exponential trend if you're in the midst of it. If you look at an exponential curve like the ones in figure 5, Kurzweil and his adherents imagine that we're at that point where the curve is increasing slowly, and it looks like incremental progress to us, but it's deceptive: the growth is about to explode.

Is the current AI spring, as many have claimed, the first harbinger of a coming explosion? Or is it simply a waypoint on a slow, incremental growth curve that won't result in human-level AI for at least another century? Or yet another AI bubble, soon to be followed by another AI winter?

To help us get some bearing on these questions, we need to take a careful look at some of the crucial abilities underlying our distinctive human intelligence, such as perception, language, decision-making, commonsense reasoning, and learning. In the next chapters, we'll see how far AI has come in capturing these abilities, and we'll assess its prospects, for 2029 and beyond.

Part II

·

Looking and Seeing

·

4

Who, What, When,

Where, Why

Look at the photo in figure 6 and tell me what you see. A woman petting a dog. A *soldier* petting a dog. A soldier who has just returned from war being welcomed by her dog, with flowers and a "Welcome Home" balloon. The soldier's face shows her complex emotions. The dog is happily wagging its tail.

When was this photo taken? Most likely within the past ten years. Where does this photo take place? Probably an airport. Why is the soldier petting the dog? She has probably been away for a long time, experienced many things, both good and bad, missed her dog a great deal, and is very happy to be home. Perhaps the dog is a symbol of all that is "home." What happened just before this photo was taken? The soldier probably got off an airplane and walked through the secure part of the airport to the place

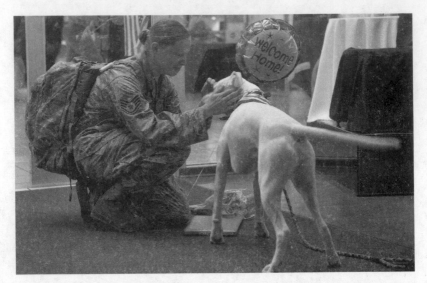

FIGURE 6: **What do you see in this photo?**

where passengers can be greeted. Her family or friends greeted her with hugs, handed her the flowers and balloon, and let go of the dog's leash. The dog came over to the soldier, who put down everything she was carrying and knelt down, carefully putting the balloon's string under her knee to keep it from floating off. What will happen next? She'll probably stand up, maybe wipe away some tears, gather her flowers, balloon, and laptop computer, take the dog's leash, and walk with the dog and her family or friends to the baggage claim area.

When you look at this picture, at the most basic level you're seeing bits of ink on a page (or pixels on a screen). Somehow your eyes and brain are able to take in this raw information and, within a few seconds, transform it into a detailed story involving living things, objects, relationships, places, emotions, motivations, and past and future actions. We look, we see, we understand. Crucially, we know what to ignore. There are many aspects of the photo that aren't strictly relevant to the story we extract from it: the pattern on the carpet, the hanging straps on the soldier's backpack, the whistle clipped to her pack's shoulder pad, the barrettes in her hair.

We humans perform this vast amount of information processing in hardly any time at all, and we have very little, if any, conscious awareness of what we're doing or how we do it. Unless you've been blind since birth, visual processing, at various levels of abstraction, dominates your brain.

Surely, the ability to describe the contents of a photograph (or a video, or a real-time stream from a camera) in this way would be one of the first things we would require for general human-level AI.

Easy Things Are Hard (Especially in Vision)

Since the 1950s, AI researchers have been trying to get computers to make sense of visual data. In the early days of AI, achieving this goal seemed relatively straightforward. In 1966, Marvin Minsky and Seymour Papert— the symbolic-AI-promoting MIT professors whom you'll recall from chapter 1—proposed the Summer Vision Project, in which they would assign undergraduates to work on "the construction of a significant part of a visual system."[1] In the words of one AI historian, "Minsky hired a first-year undergraduate and assigned him a problem to solve over the summer: connect a television camera to a computer and get the machine to describe what it sees."[2]

The undergraduate didn't get very far. And while the subfield of AI called computer vision has progressed substantially over the many decades since this summer project, a program that can look at and describe photographs in the way humans do still seems far out of reach. Vision—both looking and *seeing*—turns out to be one of the hardest of all "easy" things.

One prerequisite to describing visual input is *object recognition*— that is, recognizing a particular group of pixels in an image as a particular object category, such as "woman," "dog," "balloon," or "laptop computer." Object recognition is typically so immediate and effortless for us as humans that it didn't seem as though it would be a particularly hard problem for computers, until AI researchers actually tried to get computers to do it.

What's so hard about object recognition? Well, consider the problem of getting a computer program to recognize dogs in photographs. Figure 7 illustrates some of the difficulties. If the input is simply the pixels of the image, then the program first has to figure out which are "dog" pixels and

FIGURE 7: Object recognition: easy for humans, hard for computers

which are "non-dog" pixels (for example, background, shadows, other objects). Moreover, different dogs look very different: they can have diverse coloring, shape, and size; they can be facing in various directions; the lighting can vary considerably between images; parts of the dog can be blocked by other objects (for example, fences, people). What's more, "dog pixels" might look a lot like "cat pixels" or other animals. Under some lighting conditions, a cloud in the sky might even look very much like a dog.

Since the 1950s, the field of computer vision has struggled with these and other issues. Until recently, a major job of computer-vision researchers was to develop specialized image-processing algorithms that would identify "invariant features" of objects that could be used to recognize these objects in spite of the difficulties I sketched above. But even with sophisticated image processing, the abilities of object-recognition programs remained far below those of humans.

The Deep-Learning Revolution

The ability of machines to recognize objects in images and videos underwent a quantum leap in the 2010s due to advances in the area called deep learning.

Deep learning simply refers to methods for training "deep neural networks," which in turn refers to neural networks with more than one hidden layer. Recall that *hidden* layers are those layers of a neural network between the input and the output. The *depth* of a network is its number of hidden layers: a "shallow" network—like the one we saw in chapter 2—has only one hidden layer; a "deep" network has more than one hidden layer. It's worth emphasizing this definition: the *deep* in deep learning doesn't refer to the sophistication of what is learned; it refers only to the *depth in layers* of the network being trained.

Research on deep neural networks has been going on for several decades. What makes these networks a revolution is their recent phenomenal success in many AI tasks. Interestingly, researchers have found that the most successful deep networks are those whose structure mimics parts of the brain's visual system. The "traditional" multilayer neural networks I described in chapter 2 were inspired by the brain, but their structure is very un-brain-like. In contrast, the neural networks dominating deep learning are directly modeled after discoveries in neuroscience.

The Brain, the Neocognitron, and Convolutional Neural Networks

About the same time that Minsky and Papert were proposing their Summer Vision Project, two neuroscientists were in the midst of a decades-long study that would radically remake our understanding of vision—and particularly object recognition—in the brain. David Hubel and Torsten Wiesel were later awarded a Nobel Prize for their discoveries of hierarchical organization in the visual systems of cats and primates (including humans) and for their explanation of how the visual system transforms light striking the retina into information about what is in the scene.

Hubel and Wiesel's discoveries inspired a Japanese engineer named Kunihiko Fukushima, who in the 1970s developed one of the earliest deep neural networks, dubbed the cognitron, and its successor, the neocognitron. In his papers,[3] Fukushima reported some success training the neocognitron to recognize handwritten digits (like the ones I showed in chapter 1), but the specific learning methods he used did not seem to extend to more complex visual tasks. Nonetheless, the neocognitron was an important inspiration for later approaches to deep neural networks, including today's most influential and widely used approach: convolutional neural networks, or (as most people in the field call them) ConvNets.

ConvNets are *the* driving force behind today's deep-learning revolution in computer vision, and in other areas as well. Although they have

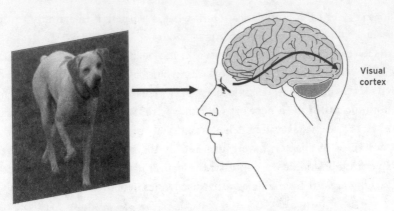

FIGURE 8: Pathway of visual input from eyes to visual cortex

been widely heralded as the next big thing in AI, ConvNets are actually not very new: they were first proposed in the 1980s by the French computer scientist Yann LeCun, who had been inspired by Fukushima's neocognitron.

I'll spend some time here describing how ConvNets work, because understanding them is crucial for making sense of where computer vision—as well as much else about AI—is today and what its limits are.

Object Recognition in the Brain and in ConvNets

Like the neocognitron, the design of ConvNets is based on several key insights about the brain's visual system that were discovered by Hubel and Wiesel in the 1950s and '60s. When your eyes focus on a scene, what they receive is light of different wavelengths that has been reflected by the objects and surfaces in the scene. Light falling on the eyes activates cells in each retina—essentially a grid of neurons in the back of the eye. These neurons communicate their activation through the optic nerves and into the brain, eventually activating neurons in the visual cortex, which resides in the back of the head (figure 8). The visual cortex is roughly organized as a hierarchical series of *layers* of neurons, like the stacked layers of a wedding cake, where the neurons in each layer communicate their activations to neurons in the succeeding layer.

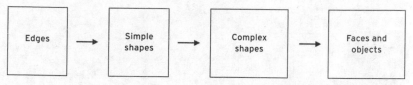

FIGURE 9: Sketch of visual features detected by neurons in different layers of the visual cortex

Hubel and Wiesel found evidence that neurons in different layers of this hierarchy act as "detectors" that respond to increasingly complex features appearing in the visual scene, as illustrated in figure 9: neurons at initial layers become active (that is, fire at a higher rate) in response to edges; their activation feeds into layers of neurons that respond to simple shapes made up of these edges; and so on, up through more complex shapes and finally entire objects and specific faces. Note that the arrows in figure 9 indicate a *bottom-up* (or *feed-forward*) flow of information, representing connections from lower to higher layers (in the figure, left to right). It's important to note that a *top-down* (or *feed-backward*) flow of information (from higher to lower layers) also occurs in the visual cortex; in fact, there are about ten times as many feed-backward connections as feed-forward ones. However, the role of these backward connections is not well understood by neuroscientists, although it is well established that our prior knowledge and expectations, presumably stored in higher brain layers, strongly influence what we perceive.

Like the feed-forward hierarchical structure illustrated in figure 9, a ConvNet consists of a sequence of layers of simulated neurons. I'll again refer to these simulated neurons as *units*. Units in each layer provide input to units in the next layer. Just like the neural network I described in chapter 2, when a ConvNet processes an image, each unit takes on a particular *activation* value—a real number that is computed from the unit's inputs and their weights.

Let's make this discussion more specific by imagining a hypothetical ConvNet, with four layers plus a "classification module," that we want to train to recognize dogs and cats in images. Assume for simplicity that each input image depicts exactly one dog or cat. Figure 10 illustrates our ConvNet's structure. It's a bit complicated, so I'll go through it carefully step-by-step to explain how it works.

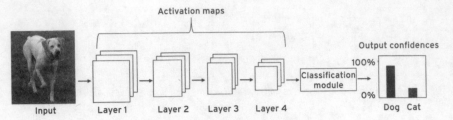

FIGURE 10: Illustration of a four-layer convolutional neural network (ConvNet) designed to recognize dogs and cats in photos

Input and Output

The input to our ConvNet is an image—that is, an array of numbers, corresponding to the brightness and color of the image's pixels.[4] Our ConvNet's final output is the network's confidence (0 percent to 100 percent) for each category: "dog" and "cat." Our goal is to have the network learn to output a high confidence for the correct category and a low confidence for the other category. In doing so, the network will learn what set of *features* of the input image are most useful for this task.

Activation Maps

Notice in figure 10 that each layer of the network is represented by a set of three overlapping rectangles. These rectangles represent *activation maps*, inspired by similar "maps" found in the brain's visual system. Hubel and Wiesel discovered that neurons in the lower layers of the visual cortex are physically arranged so that they form a rough grid, with each neuron in the grid responding to a corresponding small area of the visual field. Imagine flying at night in an airplane over Los Angeles and taking a photo; the lights seen in your photo form a rough map of the features of the lit-up city. Analogously, the activations of the neurons in each grid-like layer of the visual cortex form a rough map of the important features in the visual scene. Now imagine that you had a very special camera that could produce separate photos for house lights, building lights, and car lights. This is something like what the visual cortex does: each important visual feature has its own separate neural map. The combination of these maps is a key part of what gives rise to our perception of a scene.

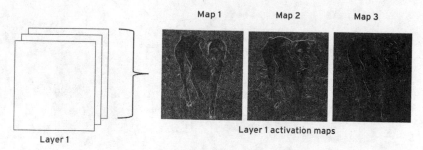

Map 1   Map 2   Map 3

Layer 1 activation maps

Layer 1

FIGURE 11: Activation maps in the first layer of our ConvNet

Like neurons in the visual cortex, the units in a ConvNet act as detectors for important visual features, each unit looking for its designated feature in a specific part of the visual field. And (very roughly) like the visual cortex, each layer in a ConvNet consists of several grids of these units, with each grid forming an activation map for a specific visual feature.

What visual features should ConvNet units detect? Let's look to the brain first. Hubel and Wiesel found that neurons in lower layers of the visual cortex act as edge detectors, where an *edge* refers to a boundary between two contrasting image regions. Each neuron receives input corresponding to a specific small region of the visual scene; this region is called the neuron's receptive field. The neuron becomes active (that is, starts firing more rapidly) only if its receptive field contains a particular kind of edge.

In fact, these neurons are quite specific about what kind of edge they respond to. Some neurons become active only when there is a vertical edge in their receptive field; some respond only to a horizontal edge; others fire only for edges at other specific angles. One of Hubel and Wiesel's most important findings was that each small region of your visual field corresponds to the receptive fields of many different such "edge detector" neurons. That is, at a low level of visual processing, your neurons are figuring out what edge orientations occur in every part of the scene you are looking at. Edge-detecting neurons feed into higher layers of the visual cortex, the neurons of which seem to be detectors for specific shapes, objects, and faces.[5]

Similarly, the first layer of our hypothetical ConvNet consists of edge-detecting units. Figure 11 gives a closer view of layer 1 of our ConvNet.

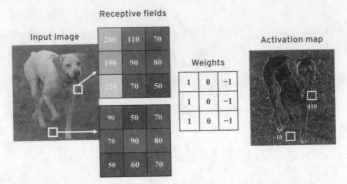

Receptive fields

Input image

Weights

1	0	−1
1	0	−1
1	0	−1

Activation map

FIGURE 12: Illustration of how convolutions are used to detect vertical edges. For example, a convolution of the upper receptive field with the weights is (200×1) + (110×0) + (70×−1) + (190×1) + (90×0) + (80×−1) + (220×1) + (70×0) + (50×−1) = 410.

This layer consists of three activation maps, each of which is a grid of units. Each unit in a map corresponds to the analogous location in the input image, and each unit gets its input from a small region around that location—its receptive field. (The receptive fields of neighboring units typically overlap.) Each unit in each map calculates an activation value that measures the degree to which the region matches the unit's preferred edge orientation—for example, vertical, horizontal, or slanted at various degrees.

Figure 12 illustrates in detail how the units in map 1—those that detect vertical edges—calculate their activations. The small white squares in the input image represent the receptive fields of two different units. The image patches inside these receptive fields, when enlarged, are shown as arrays of pixel values. Here, for simplicity, I've displayed each patch as a three-by-three set of pixels (the values, by convention, range from 0 to 255—the lighter the pixel, the higher the value). Each unit receives as input the pixel values in its receptive field. The unit then multiplies each input by its weight and sums the results to produce the unit's activation.

The weights shown in figure 12 are designed to produce a high positive activation when there is a light-to-dark vertical edge in the receptive field (that is, high contrast between the left and the right sides of the input patch). The upper receptive field contains a vertical edge: the dog's light fur next to the darker grass. This is reflected in the high activation value (410). The lower receptive field does not contain such an edge, only dark grass, and the activation (−10) is closer to 0. Note that a dark-to-light vertical edge will yield a "high" negative value (that is, a negative value far from 0).

This calculation—multiplying each value in a receptive field by its corresponding weight and summing the results—is called a convolution. Hence the name "convolutional neural network." I mentioned above that in a ConvNet, an activation map is a grid of units corresponding to receptive fields all over the image. Each unit in a given activation map uses the same weights to compute a convolution with its receptive field; imagine the input image with the white square sliding along every patch of the image.[6] The result is the activation map in figure 12: the center pixel of a unit's receptive field is colored white for high positive and negative activations and darker for activations close to 0. You can see that the white areas highlight the locations where vertical edges exist. Maps 2 and 3 in figure 11 were created in the same way, but with weights that highlight horizontal and slanted edges, respectively. Taken together, the maps of edge-detecting units in layer 1 provide the ConvNet with a representation of the input image in terms of oriented edges in different regions, something like what an edge-detection program would produce.

Let's take a moment to talk about the word *map* here. In everyday use, *map* refers to a spatial representation of a geographic area, such as a city. A road map of Paris, say, shows a particular *feature* of the city—its layout of streets, avenues, and alleys—but doesn't include the city's many other features, such as buildings, houses, lampposts, trash cans, apple trees, and fishponds. Other kinds of maps focus on other features; you can find maps that highlight Paris's bike lanes, its vegetarian restaurants, its dog-friendly parks. Whatever your interests, there is quite possibly a map that shows where to find them. If you wanted to explain Paris to a friend who had never been there, a creative approach might be to show your friend a collection of such "special interest" maps of the city.

A ConvNet (like the brain) represents the visual scene as a collection of maps, reflecting the specific "interests" of a set of detectors. In my example in figure 11, these interests are different edge orientations. However, as we'll see below, in ConvNets the network itself *learns* what its interests (that is, detectors) should be; these depend on the specific task it is trained for.

Making maps isn't limited to layer 1 of our ConvNet. As you can see in figure 10, a similar structure applies at all of the layers: each layer has a set of detectors, each of which creates its own activation map. A key to the ConvNet's success is that—again, inspired by the brain—these maps are *hierarchical*: the inputs to the units at layer 2 are the activation maps of layer 1, the inputs to the units at layer 3 are the activation maps of layer 2, and so on up the layers. In our hypothetical network, in which layer 1 units respond to edges, the layer 2 units would be sensitive to specific combinations of edges, such as corners and T shapes. Layer 3 detectors would be sensitive to combinations of combinations of edges. As you go up the hierarchy, the detectors become sensitive to increasingly more complex features, just as Hubel, Wiesel, and others saw in the brain.

Our hypothetical ConvNet has four layers, each with three maps, but in the real world these networks can have many more layers—sometimes hundreds—each with different numbers of activation maps. Determining these and many other aspects of a ConvNet's structure is part of the art of getting these complex networks to work for a given task. In chapter 3, I described I. J. Good's vision of a future "intelligence explosion" in which machines themselves create increasingly intelligent machines. We're not there yet. For the time being, getting ConvNets to work well requires a lot of human ingenuity.

Classification in ConvNets

Layers 1 to 4 of our network are called convolutional layers because each performs convolutions on the preceding layer (and layer 1 performs convolutions on the input). Given an input image, each layer successively performs its calculations, and finally at layer 4 the network has produced a set of activation maps for relatively complex features. These might include eyes, leg shapes, tail shapes, or anything else that the network has learned is useful for classifying the objects it is trained on (here dogs and cats). At this point, it's time for the classification module to use these features to predict what object the image depicts.

The classification module is actually an entire traditional neural network, similar to the kind I described in chapter 2.[7] The inputs to the classification module are the activation maps from the highest convolutional layer. The module's output is a set of percentage values, one for each possible category, rating the network's confidence that the input depicts an image of that category (here dog or cat).

Let me summarize this brief explanation of ConvNets: Inspired by Hubel and Wiesel's findings on the brain's visual cortex, a ConvNet takes an input image and transforms it—via convolutions—into a set of activation maps with increasingly complex features. The features at the highest convolutional layer are fed into a traditional neural network (which I've called the classification module), which outputs confidence percentages for the network's known object categories. The object category with the highest confidence is returned as the network's classification of the image.[8]

Would you like to experiment with a well-trained ConvNet? Simply take a photo of an object, and upload it to Google's "search by image" engine.[9] Google will run a ConvNet on your image and, based on the resulting confidences (over thousands of possible object categories), will tell you its "best guess" for the image.

Training a ConvNet

Our hypothetical ConvNet consists of edge detectors at its first layer, but in real-world ConvNets edge detectors aren't built in. Instead, ConvNets learn from training examples what features should be detected at each layer, as well as how to set the weights in the classification module so as to produce a high confidence for the correct answer. And, just as in traditional neural networks, all the weights can be learned from data via the same back-propagation algorithm that I described in chapter 2.

More specifically, here is how you could train our ConvNet to identify a given image as a dog or cat. First, collect many example images of dogs and cats—this is your "training set." Also, create a file that gives a label for each image—that is, "dog" or "cat." (Or better, take a hint from computer-vision researchers: Hire a graduate student to do all this for you. If you are a graduate student, then recruit an undergrad. No one enjoys this labeling chore!) Your training program initially sets all the weights in the network to random values. Then your program commences training: one by one, each image is given as the input to the network; the network performs its layer-by-layer calculations and finally outputs confidence percentages for "dog" and "cat." For each image, your training program compares these output values to the "correct" values; for example, if the image is a dog, then "dog" confidence should be 100 percent and "cat" confidence should be 0 percent. Then the training program uses the back-propagation algorithm to change the weights throughout the network just a bit, so that the next time this image is seen, the confidences will be closer to the correct values.

Following this procedure—input the image, then calculate the error at the output, then change the weights—for every image in your training set is called one "epoch" of training. Training a ConvNet requires many epochs, during which the network processes each image over and over again. Initially, the network will be very bad at recognizing dogs and cats, but slowly, as it changes its weights over many epochs, it will get increasingly better at the task. Finally, at some point, the network "converges"; that is, the weights stop changing much from one epoch to the next, and the network is (in principle!) very good at recognizing dogs and cats in the images in the training set. But we won't know if the network is actually good at this task in general until we see if it can apply what it has learned to identify images from outside its training set. What's really interesting is that, even though ConvNets are not constrained by a programmer to learn to detect any particular feature, when trained on large sets of real-world photographs, they indeed seem to learn a hierarchy of detectors similar to what Hubel and Wiesel found in the brain's visual system.

In the next chapter, I'll recount the extraordinary ascent of ConvNets from relative obscurity to near-complete dominance in machine vision, a transformation made possible by a concurrent technological revolution: that of "big data."

5

ConvNets and

ImageNet

Yann LeCun, the inventor of ConvNets, has worked on neural networks all of his professional life, starting in the 1980s and continuing through the winters and springs of the field. As a graduate student and postdoctoral fellow, he was fascinated by Rosenblatt's perceptrons and Fukushima's neocognitron, but noted that the latter lacked a good supervised-learning algorithm. Along with other researchers (most notably, his postdoctoral advisor Geoffrey Hinton), LeCun helped develop such a learning method—essentially the same form of back-propagation used on ConvNets today.[1]

In the 1980s and '90s, while working at Bell Labs, LeCun turned to the problem of recognizing handwritten digits and letters. He combined ideas from the neocognitron with the back-propagation algorithm to create the semi-eponymous "LeNet"—one of the earliest ConvNets. LeNet's handwritten-digit-recognition abilities made it a commercial success: in the 1990s and into the 2000s it was used by the U.S. Postal Service for automated zip code recognition, as well as in the banking industry for automated reading of digits on checks.

LeNet and its successor ConvNets did not do well in scaling up to more complex vision tasks. By the mid-1990s, neural networks started falling out of favor in the AI community, and other methods came to dominate the field. But LeCun, still a believer, kept working on ConvNets, gradually improving them. As Geoffrey Hinton later said of LeCun, "He kind of carried the torch through the dark ages."[2]

LeCun, Hinton, and other neural network loyalists believed that improved, larger versions of ConvNets and other deep networks would conquer computer vision if only they could be trained with enough data. Stubbornly, they kept working on the sidelines throughout the 2000s. In 2012, the torch carried by ConvNet researchers suddenly lit the vision world afire, by winning a computer-vision competition on an image data set called ImageNet.

Building ImageNet

AI researchers are a competitive bunch, so it's no surprise that they like to organize competitions to drive the field forward. In the field of visual object recognition, researchers have long held annual contests to determine whose program performs the best. Each of these contests features a "benchmark data set": a collection of photos, along with human-created labels that name objects in the photos.

From 2005 to 2010, the most prominent of these annual contests was the PASCAL Visual Object Classes competition, which by 2010 featured about fifteen thousand photographs (downloaded from the photo-sharing site Flickr), with human-created labels for twenty object categories, such as "person," "dog," "horse," "sheep," "car," "bicycle," "sofa," and "potted plant."

The entries to the "classification" part of this contest[3] were computer-vision programs that could take a photograph as input (without seeing its human-created label) and could then output, for each of the twenty categories, whether an object of that category was present in the image.

Here's how the competition worked. The organizers would split the photographs into a training set that contestants could use to train their programs and a test set, not released to contestants, that would be used to gauge the programs' performance on images outside the training set. Prior to the competition, the training set would be offered online, and when the contest was held, researchers would submit their trained programs to be tested on the secret test set. The winning entry was the one that had the highest accuracy recognizing objects in the test-set images.

The annual PASCAL competitions were a very big deal and did a lot to spur research in object recognition. Over the years of the challenge, the competing programs gradually got better (curiously, potted plants remained the hardest objects to recognize). However, some researchers were frustrated by the shortcomings of the PASCAL benchmark as a way to move computer vision forward. Contestants were focusing too much on PASCAL's specific twenty object categories and were not building systems that could scale up to the huge number of object categories recognized by humans. Furthermore, there just weren't enough photos in the data set for the competing systems to learn all the many possible variations in what the objects look like so as to be able to generalize well.

To move ahead, the field needed a new benchmark image collection, one featuring a much larger set of categories and vastly more photos. Fei-Fei Li, a young computer-vision professor at Princeton, was particularly focused on this goal. By serendipity, she learned of a project led by a fellow Princeton professor, the psychologist George Miller, to create a database of English words, arranged in a hierarchy moving from most specific to most general, with groupings among synonyms. For example, consider the word *cappuccino*. The database, called WordNet, contains the following information about this term (where an arrow means "is a kind of"):

cappuccino ⇒ *coffee* ⇒ *beverage* ⇒ *food* ⇒ *substance* ⇒ *physical entity* ⇒ *entity*

The database also contains information that, say, *beverage*, *drink*, and *potable* are synonyms, that *beverage* is part of another chain including *liquid*, and so forth.

WordNet had been (and continues to be) used extensively in research by psychologists and linguists as well as in AI natural-language processing systems, but Fei-Fei Li had a new idea: create an image database that is structured according to the nouns in WordNet, where each noun is linked to a large number of images containing examples of that noun. Thus the idea for ImageNet was born.

Li and her collaborators soon commenced collecting a deluge of images by using WordNet nouns as queries on image search engines such as Flickr and Google image search. However, if you've ever used an image search engine, you know that the results of a query are often far from perfect. For example, if you type "macintosh apple" into Google image search, you get photos not only of apples and Mac computers but also of apple-shaped candles, smartphones, bottles of apple wine, and any number of other nonrelevant items. Thus, Li and her colleagues had to have humans figure out which images were not actually illustrations of a given noun and get rid of them. At first, the humans who did this were mainly undergraduates. The work was agonizingly slow and taxing. Li soon figured out that at the rate they were going, it would take ninety years to complete the task.[4]

Li and her collaborators brainstormed about possible ways to automate this work, but of course the problem of deciding if a photo is an instance of a particular noun *is* the task of object recognition itself! And computers were nowhere near to being reliable at this task, which was the whole reason for constructing ImageNet in the first place.

The group was at an impasse, until Li, by chance, stumbled upon a three-year-old website that could deliver the human smarts that ImageNet required. The website had the strange name Amazon Mechanical Turk.

Mechanical Turk

According to Amazon, its Mechanical Turk service is "a marketplace for work that requires human intelligence." The service connects *requesters*, people who need a task accomplished that is hard for computers, with *workers*, people who are willing to lend their human intelligence to a requester's task, for a small fee (for example, labeling the objects in a photo, for ten cents per photo). Hundreds of thousands of workers have signed up, from all over the world. Mechanical Turk is the embodiment of Marvin Minsky's "Easy things are hard" dictum: the human workers are hired to perform the "easy" tasks that are currently too hard for computers.

The name Mechanical Turk comes from a famous eighteenth-century AI hoax: the original Mechanical Turk was a chess-playing "intelligent machine," which secretly hid a human who controlled a puppet (the "Turk," dressed like an Ottoman sultan) that made the moves. Evidently, it fooled many prominent people of the time, including Napoleon Bonaparte. Amazon's service, while not meant to fool anyone, is, like the original Mechanical Turk, *"Artificial* Artificial Intelligence."[5]

Fei-Fei Li realized that if her group paid tens of thousands of workers on Mechanical Turk to sort out irrelevant images for each of the WordNet terms, the whole data set could be completed within a few years at a relatively low cost. In a mere two years, more than three million images were labeled with corresponding WordNet nouns to form the ImageNet data set. For the ImageNet project, Mechanical Turk was "a godsend."[6] The service continues to be widely used by AI researchers for creating data sets; nowadays, academic grant proposals in AI commonly include a line item for "Mechanical Turk workers."

The ImageNet Competitions

In 2010, the ImageNet project launched the first ImageNet Large Scale Visual Recognition Challenge, in order to spur progress toward more general object-recognition algorithms. Thirty-five programs competed, representing computer-vision researchers from academia and industry around the world. The competitors were given labeled training images—1.2 million of them—and a list of possible categories. The task for the trained programs was to output the correct category of each input image. The ImageNet competition had a thousand possible categories, compared with PASCAL's twenty.

The thousand possible categories were a subset of WordNet terms chosen by the organizers. The categories are a random-looking assembly of nouns, ranging from the familiar and commonplace ("lemon," "castle," "grand piano") to the somewhat less common ("viaduct," "hermit crab," "metronome"), and on to the downright obscure ("Scottish deerhound," "ruddy turnstone," "hussar monkey"). In fact, obscure animals and plants—at least ones I wouldn't be able to distinguish—constitute at least a tenth of the thousand target categories.

Some of the photographs contain only one object; others contain many objects, including the "correct" one. Because of this ambiguity, a program gets to guess five categories for each image, and if the correct one is in this list, the program is said to be correct on this image. This is called the "top-5" accuracy metric.

The highest-scoring program in 2010 used a so-called support vector machine, the predominant object-recognition algorithm of the day, which employed sophisticated mathematics to learn how to assign a category to each input image. Using the top-5 accuracy metric, this winning program was correct on 72 percent of the 150,000 test images. Not a bad showing, though this means that the program was wrong, even with five guesses allowed, on more than 40,000 of the test images, leaving a lot of room for improvement. Notably, there were no neural networks among the top-scoring programs.

The following year, the highest-scoring program—also using support vector machines—showed a respectable but modest improvement, getting 74 percent of the test images correct. Most people in the field expected this trend to continue; computer-vision research would chip away at the problem, with gradual improvement at each annual competition.

However, these expectations were upended in the 2012 ImageNet competition: the winning entry achieved an amazing 85 percent correct. Such a jump in accuracy was a shocking development. What's more, the winning entry did not use support vector machines or any of the other dominant computer-vision methods of the day. Instead, it was a convolutional neural network. This particular ConvNet has come to be known as AlexNet, named after its main creator, Alex Krizhevsky, then a graduate student at the University of Toronto, supervised by the eminent neural network researcher Geoffrey Hinton. Krizhevsky, working with Hinton and a fellow student, Ilya Sutskever, created a scaled-up version of Yann LeCun's LeNet from the 1990s; training such a large network was now made possible by increases in computer power. AlexNet had eight layers, with about sixty million weights whose values were learned via back-propagation from the million-plus training images.[7] The Toronto group came up with some clever methods for making the network training work better, and it took a cluster of powerful computers about a week to train AlexNet.

AlexNet's success sent a jolt through the computer-vision and broader AI communities, suddenly waking people up to the potential power of ConvNets, which most AI researchers hadn't considered a serious con-tender in modern computer vision. In a 2015 article, the journalist Tom Simonite interviewed Yann LeCun about the unexpected triumph of ConvNets:

> LeCun recalls seeing the community that had mostly ignored neural networks pack into the room where the winners presented a paper on their results. "You could see right there a lot of senior people in the com-munity just flipped," he says. "They said, 'Okay, now we buy it. That's it, now—you won.'"[8]

At almost the same time, Geoffrey Hinton's group was also demon-strating that deep neural networks, trained on huge amounts of labeled data, were significantly better than the current state of the art in speech recognition. The Toronto group's ImageNet and speech-recognition results had substantial ripple effects. Within a year, a small company started by Hinton was acquired by Google, and Hinton and his students Krizhevsky and Sutskever became Google employees. This acqui-hire in-stantly put Google at the forefront of deep learning.

Soon after, Yann LeCun was lured away from his full-time New York University professorship by Facebook to head up its newly formed AI lab. It didn't take long before all the big tech companies (as well as many smaller ones) were snapping up deep-learning experts and their graduate students as fast as possible. Seemingly overnight, deep learning became the hottest part of AI, and expertise in deep learning guaranteed computer scientists a large salary in Silicon Valley or, better yet, venture capital funding for their proliferating deep-learning start-up companies.

The annual ImageNet competition began to see wider coverage in the media, and it quickly morphed from a friendly academic contest into a high-profile sparring match for tech companies commercializing computer vision. Winning at ImageNet would guarantee coveted respect from the vision community, along with free publicity, which might translate into product sales and higher stock prices. The pressure to produce programs that outperformed competitors was notably manifest in a 2015 cheating incident involving the giant Chinese internet company Baidu. The cheating involved a subtle example of what people in machine learning call data snooping.

Here's what happened: Before the competition, each team competing on ImageNet was given training images labeled with correct object categories. They were also given a large test set—a collection of images not in the training set—without any labels. Once a program was trained, a team could see how well their method performed on this test set. This helps test how well a program has learned to generalize (as opposed to, say, memorizing the training images and their labels). Only the performance on the test set counts. The way a team could find out how well their program did on the test set was to run their program on each test-set image, collect the top five guesses for each image, and submit this list to a "test server"—a computer run by the contest organizers. The test server would compare the submitted list with the (secret) correct answers and spit out the percentage correct.

Each team could sign up for an account on the test server and use it to see how well various versions of their programs were scoring; this would allow them to publish (and publicize) their results before the official results were announced.

A cardinal rule in machine learning is "Don't train on the test data." It seems obvious: If you include test data in any part of training your program, you won't get a good measure of the program's generalization abilities. It would be like giving students the questions on the final exam before they take the test. But it turns out that there are subtle ways that this rule can be unintentionally (or intentionally) broken to make your program's performance look better than it actually is.

One such method would be to submit your program's test-set answers to the test server and, based on the result, tweak your program. Then submit again. Repeat this many times, until you have tweaked it to do better on the test set. This doesn't require seeing the actual labels in the test set, but it does require getting feedback on accuracy and adjusting your program accordingly. It turns out that if you can do this enough times, it can be very effective in improving your program's performance on the test set. But because you're using information from the test set to change your program, you've now destroyed the ability to use the test set to see if your program generalizes well. It would be like allowing students to take a final exam many times, each time getting back a single grade, but using that single grade to try to improve their performance the next time around. Then, at the end, the students submit the version of their answers that got them the best score. This is no longer a good measure of how well the students have learned the subject, just a measure of how they adapted their answers to particular test questions.

To prevent this kind of data snooping while still allowing the ImageNet competitors to see how well their programs are doing, the organizers set a rule saying that each team could submit answers to the test server at most twice per week. This would limit the amount of feedback the teams could glean from the test runs.

The great ImageNet battle of 2015 was fought over a fraction of a percentage point—seemingly trivial but potentially very lucrative. Early in the year, a team from Baidu announced a method that achieved the highest (top-5) accuracy yet on an ImageNet test set: 94.67 percent, to be exact. But on the very same day, a team from Microsoft announced a better accuracy with their method: 95.06 percent. A few days later, a rival team from Google announced a slightly different method that did even better: 95.18 percent. This record held for a few months, until Baidu made a new announcement: it had improved its method and now could boast a new record, 95.42 percent. This result was widely publicized by Baidu's public relations team.

But within a few weeks, a terse announcement came from the ImageNet organizers: "During the period of November 28th, 2014 to May 13th, 2015, there were at least 30 accounts used by a team from Baidu to submit to the test server at least 200 times, far exceeding the specified limit of two submissions per week."[9] In short, the Baidu team had been caught data snooping.

The two hundred points of feedback potentially allowed the Baidu team to determine which tweaks to their program would make it perform best on this test set, gaining it the all-important fraction of a percentage point that made the win. As punishment, Baidu was disqualified from entering its program in the 2015 competition.

Baidu, hoping to minimize bad publicity, promptly apologized and then laid the blame on a rogue employee: "We found that a team leader had directed junior engineers to submit more than two submissions per week, a breach of the current ImageNet rules."[10] The employee, though disputing that he had broken any rules, was promptly fired from the company.

While this story is merely an interesting footnote to the larger history of deep learning in computer vision, I tell it to illustrate the extent to which the ImageNet competition came to be seen as *the* key symbol of progress in computer vision, and AI in general.

Cheating aside, progress on ImageNet continued. The final competition was held in 2017, with a winning top-5 accuracy of 98 percent. As one journalist commented, "Today, many consider ImageNet solved,"[11] at least for the classification task. The community is moving on to new benchmark data sets and new problems, especially ones that integrate vision and language.

What was it that enabled ConvNets, which seemed to be at a dead end in the 1990s, to suddenly dominate the ImageNet competition, and subsequently most of computer vision in the last half a decade? It turns out that the recent success of deep learning is due less to new breakthroughs in AI than to the availability of huge amounts of data (thank you, internet!) and very fast parallel computer hardware. These factors, along with improvements in training methods, allow hundred-plus-layer networks to be trained on millions of images in just a few days.

Yann LeCun himself was taken by surprise at how fast things turned around for his ConvNets: "It's rarely the case where a technology that has been around for 20, 25 years—basically unchanged—turns out to be the best. The speed at which people have embraced it is nothing short of amazing. I've never seen anything like this before."[12]

The ConvNet Gold Rush

Once ImageNet and other large data sets gave ConvNets the vast amount of training examples they needed to work well, companies were suddenly able to apply computer vision in ways never seen before. As Google's Blaise Agüera y Arcas remarked, "It's been a sort of gold rush—attacking one problem after another with the same set of techniques."[13] Using ConvNets trained with deep learning, image search engines offered by Google, Microsoft, and others were able to vastly improve their "find similar images" feature. Google offered a photo-storage system that would tag your photos by describing the objects they contained, and Google's Street View service could recognize and blur out street addresses and license plates in its images. A proliferation of mobile apps enabled smartphones to perform object and face recognition in real time.

Facebook labeled your uploaded photos with names of your friends and registered a patent on classifying the emotions behind facial expressions in uploaded photos; Twitter developed a filter that could screen tweets for pornographic images; and several photo- and video-sharing sites started applying tools to detect imagery associated with terrorist groups. ConvNets can be applied to video and used in self-driving cars to track pedestrians, or to read lips and classify body language. ConvNets can even diagnose breast and skin cancer from medical images, determine the stage of diabetic retinopathy, and assist physicians in treatment planning for prostate cancer.

These are just a few examples of the many existing (or soon-to-exist) commercial applications powered by ConvNets. In fact, there's a good chance that any modern computer-vision application you use employs ConvNets. Moreover, there's an excellent chance it was "pretrained" on images from ImageNet to learn generic visual features before being "fine-tuned" for more specific tasks.

Given that the extensive training required by ConvNets is feasible only with specialized computer hardware—typically, powerful graphical processing units (GPUs)—it is not surprising that the stock price of the NVIDIA Corporation, the most prominent maker of GPUs, increased by over 1,000 percent between 2012 and 2017.

Have ConvNets Surpassed Humans
at Object Recognition?

As I learned more about the remarkable success of ConvNets, I wondered how close they were to rivaling our own human object-recognition abilities. A 2015 paper from Baidu (post–cheating scandal) carried the subtitle "Surpassing Human-Level Performance on ImageNet Classification."[14] At about the same time, Microsoft announced in a research blog "a major advance in technology designed to identify the objects in a photograph or video, showcasing a system whose accuracy meets and sometimes exceeds human-level performance."[15] While both companies made it clear they were talking about accuracy specifically on ImageNet, the media were not so careful, giving way to sensational headlines such as "Computers Now Better than Humans at Recognising and Sorting Images" and "Microsoft Has Developed a Computer System That Can Identify Objects Better than Humans."[16]

Let's look a bit harder at the specific contention that machines are now "better than humans" at object recognition on ImageNet. This assertion is based on a claim that humans have an error rate of about 5 percent, whereas the error rate of machines is (at the time of this writing) close to 2 percent. Doesn't this confirm that machines are better than humans at this task? As is often the case for highly publicized claims about AI, the claim comes with a few caveats.

Here's one caveat. When you read about a machine "identifying objects correctly," you'd think that, say, given an image of a basketball, the machine would output "basketball." But of course, on ImageNet, correct identification means only that the correct category is in the machine's top-five categories. If, given an image of a basketball, the machine outputs "croquet ball," "bikini," "warthog," "basketball," and "moving van," in that order, it is considered correct. I don't know how often this kind of thing happens, but it's notable that the best *top-1 accuracy*—the fraction of test images on which the correct category is at the top of the list—was about 82 percent, compared with 98 percent top-5 accuracy, in the 2017 ImageNet competition. No one, as far as I know, has reported a comparison between machines and humans on top-1 accuracy.

Here's another caveat. Consider the claim, "Humans have an error rate of about 5% on ImageNet." It turns out that saying "humans" is not quite accurate; this result is from an experiment involving a *single human*, one Andrej Karpathy, who was at the time a graduate student at Stanford, researching deep learning. Karpathy wanted to see if he could train himself to compete against the best ConvNets on ImageNet. Considering that ConvNets train on 1.2 million images and then are run on 150,000 test images, this is a daunting task for a human. Karpathy, who has a popular blog about AI, wrote about his experience:

> I ended up training [myself] on 500 images and then switched to [a reduced] test set of 1,500 images. The labeling [that is, Karpathy's guessing five categories per image] happened at a rate of about 1 per minute, but this decreased over time. I only enjoyed the first ~200, and the rest I only did #forscience. . . . Some images are easily recognized, while some images (such as those of fine-grained breeds of dogs, birds, or monkeys) can require multiple minutes of concentrated effort. I became very good at identifying breeds of dogs.[17]

Karpathy found that he was wrong on about 75 of his 1,500 test images, and he went on to analyze the errors he made, finding that they were largely due to images with multiple objects, images with specific breeds of dogs, species of birds or plants, and so on, and object categories that he didn't realize were included in the target categories. The kinds of errors made by ConvNets are different: while they also get confused by images containing multiple objects, unlike humans they tend to miss objects that are small in the image, objects that have been distorted by color or contrast filters the photographer applied to the image, and "abstract representations" of objects, such as a painting or statue of a dog, or a stuffed toy dog. Thus, the claim that computers have bested humans on ImageNet needs to be taken with a large grain of salt.

Here's a caveat that might surprise you. When a human says that a photo contains, say, a dog, we assume it's because the human actually saw a dog in the photo. But if a ConvNet correctly says "dog," how do we know it actually is basing this classification on the dog in the image? Maybe there's something else in the image—a tennis ball, a Frisbee, a chewed-up shoe—that was often associated with dogs in the training images, and the ConvNet is recognizing these and assuming there is a dog in the photo. These kinds of correlations have often ended up fooling machines.

One thing we could do is ask the machine to not only output an object category for an image but also learn to draw a box around the target object, so we know the machine has actually "seen" the object. This is precisely what the ImageNet competition started doing in its second year with its "localization challenge." The localization task provided training images with such boxes drawn (by Mechanical Turk workers) around the target object(s) in each image; on the test images, the task for competing programs was to predict five object categories each with the coordinates of a corresponding box. What may be surprising is that while deep convolutional neural networks have performed very well at localization, their performance has remained significantly worse than their performance on categorization, although newer competitions are focusing on precisely this problem.

Probably the most important differences between today's ConvNets and humans when it comes to recognizing objects are in how learning takes place and in how robust and reliable that learning turns out to be. I'll explore these differences in the next chapter.

The caveats I described above aren't meant to diminish the amazing recent progress in computer vision. There is no question that convolutional neural networks have been stunningly successful in this and other areas, and these successes have not only produced commercial products but also resulted in a real sense of optimism in the AI community. My discussion is meant to illustrate how challenging vision turns out to be and to add some perspective on the progress made so far. Object recognition is not yet close to being "solved" by artificial intelligence.

Beyond Object Recognition

I have focused on object recognition in this chapter because this has been the area in which computer vision has recently seen the most progress. However, there's obviously a lot more to vision than just recognizing objects. If the goal of computer vision is to "get a machine to describe what it sees," then machines will need to recognize not only objects but also their relationships to one another and how they interact with the world. If the "objects" in question are living beings, the machines will need to know something about their actions, goals, emotions, likely next steps, and all the other aspects that figure into telling the story of a visual scene. Moreover, if we really want the machines to *describe* what they see, they will need to use language. AI researchers are actively working on getting machines to do these things, but as usual these "easy" things are very hard. As the computer-vision expert Ali Farhadi told *The New York Times*, "We're still very, very far from visual intelligence, understanding scenes and actions the way humans do."[18]

Why are we still so far from this goal? It seems that visual intelligence isn't easily separable from the rest of intelligence, especially general knowledge, abstraction, and language—abilities that, interestingly, involve parts of the brain that have many feedback connections to the visual cortex. Additionally, it could be that the knowledge needed for humanlike visual intelligence—for example, making sense of the "soldier and dog" photo at the beginning of the previous chapter—can't be learned from millions of pictures downloaded from the web, but has to be experienced in some way in the real world.

In the next chapter, I'll look more closely at machine learning in vision, focusing in particular on the differences between the ways humans and machines learn and trying to tease out just what the machines we have trained have actually learned.

6

·

A Closer Look at

·

Machines That Learn

·

The deep-learning pioneer Yann LeCun has received many awards and accolades, but perhaps his ultimate (if geeky) honor is being the subject of a widely followed and very funny parody Twitter account sporting the name "Bored Yann LeCun." With the description "Musing on the rise of deep learning in Yann's downtime," the anonymously authored account frequently ends its clever in-joke tweets with the hashtag #FeelTheLearn.[1]

Indeed, media reports on cutting-edge AI have been "feeling the learn" by celebrating the power of deep learning—emphasis on "learning." We are told, for example, that "we can now build systems that learn how to perform tasks on their own,"[2] that "deep learning [enables] computers to literally teach themselves,"[3] and that deep-learning systems learn "in a way similar to the human brain."[4]

In this chapter, I'll look in more detail at how machines—particularly ConvNets—learn and how their learning processes contrast with those of humans. Furthermore, I'll explore how differences between learning in ConvNets and in humans affect the robustness and trustworthiness of what is learned.

Learning on One's Own

The learning-from-data approach of deep neural networks has generally proved to be more successful than the "good old-fashioned AI" strategy, in which human programmers construct explicit rules for intelligent behavior. However, contrary to what some media have reported, the learning process of ConvNets is not very humanlike.

As we've seen, the most successful ConvNets learn via a *supervised-learning* procedure: they gradually change their weights as they process the examples in the training set again and again, over many epochs (that is, many passes through the training set), learning to classify each input as one of a fixed set of possible output categories. In contrast, even the youngest children learn an open-ended set of categories and can recognize instances of most categories after seeing only a few examples. Moreover, children don't learn passively: they ask questions, they demand information on the things they are curious about, they infer abstractions of and connections between concepts, and, above all, they actively explore the world.

It is inaccurate to say that today's successful ConvNets learn "on their own." As we saw in the previous chapter, in order for a ConvNet to learn to perform a task, a huge amount of human effort is required to collect, curate, and label the data, as well as to design the many aspects of the ConvNet's architecture. While ConvNets use back-propagation to learn their "parameters" (that is, weights) from training examples, this learning is enabled by a collection of what are called "hyperparameters"—an umbrella term that refers to all the aspects of the network that need to be set up by humans to allow learning to even begin. Examples of hyperparameters include the number of layers in the network, the size of the units' "receptive fields" at each layer, how large the change in each weight should be during learning (called the learning rate), and many other technical details of the training process. This part of setting up a ConvNet is called tuning the hyperparameters. There are many values to set as well as complex design decisions to be made, and these settings and designs interact with one another in complex ways to affect the ultimate performance of the network. Moreover, these settings and designs must typically be decided anew for each task a network is trained on.

Tuning the hyperparameters might sound like a pretty mundane activity, but doing it well is absolutely crucial to the success of ConvNets and other machine-learning systems. Because of the open-ended nature of designing these networks, in general it is not possible to automatically set all the parameters and designs, even with automated search. Often it takes a kind of cabalistic knowledge that students of machine learning gain both from their apprenticeships with experts and from hard-won experience. As Eric Horvitz, director of Microsoft's research lab, characterized it, "Right now, what we are doing is not a science but a kind of alchemy."[5] And the people who can do this kind of "network whispering" form a small, exclusive club: according to Demis Hassabis, cofounder of Google DeepMind, "It's almost like an art form to get the best out of these systems. . . . There's only a few hundred people in the world that can do that really well."[6]

Actually, the number of deep-learning experts is growing quickly; many universities now offer courses in the subject, and a growing list of companies have started their own deep-learning training programs for employees. Membership in the deep-learning club can be quite lucrative. At a recent conference I attended, a leader of Microsoft's AI product group spoke to the audience about the company's efforts to hire young deep-learning engineers: "If a kid knows how to train five layers of neural networks, the kid can demand five figures. If the kid knows how to train fifty layers, the kid can demand seven figures."[7] Lucky for this soon-to-be-wealthy kid, the networks can't yet teach themselves.

Big Data

It's no secret: deep learning requires *big data*. Big in the sense of the million-plus labeled training images in ImageNet. Where does all this data come from? The answer is, of course, you—and probably everyone you know.

Modern computer-vision applications are possible only because of the billions of images that internet users have uploaded and (sometimes) tagged with text identifying what is in the image. Have you ever put a photo of a friend on your Facebook page and commented on it? Facebook thanks you! That image and text might have been used to train its face-recognition system. Have you ever uploaded an image to Flickr? If so, it's possible your image is part of the ImageNet training set. Have you ever identified a picture in order to prove to a website that you're not a robot? Your identification might have helped Google tag an image for use in training its image search system.

Big tech companies offer many services for free on your computer and smartphone: web search, video calling, email, social networking, automated personal assistants—the list goes on. What's in it for these companies? The answer you might have heard is that their true product is their *users* (like you and me); their *customers* are the advertisers who grab our attention and information about us while we use these "free" services. But there's a second answer: when we use services provided by tech companies such as Google, Amazon, and Facebook, we are directly providing these companies with examples—in the form of our images, videos, text, or speech—that they can utilize to better train their AI programs. And these improved programs attract more users (and thus more data), helping advertisers to target their ads more effectively. Moreover, the training examples we provide them can be used to train and offer services such as computer vision and natural-language processing to businesses for a fee.

Much has been written about the ethics of these big companies using data you have created (such as all the images, videos, and text that you upload to Facebook) to train programs and sell products without informing or compensating you. This is an important discussion but beyond the scope of this book.[8] The point I want to make here is that the reliance on extensive collections of labeled training data is one more way in which deep learning differs from human learning.

With the proliferation of deep-learning systems in real-world applications, companies are finding themselves in need of new labeled data sets for training deep neural networks. Self-driving cars are a noteworthy example.

These cars need sophisticated computer vision in order to recognize lanes in the road, traffic lights, stop signs, and so on, and to distinguish and track different kinds of potential obstacles, such as other cars, pedestrians, bicyclists, animals, traffic cones, knocked-over garbage cans, tumbleweeds, and anything else that you might not want your car to hit. Self-driving cars need to learn what these various objects look like—in sun, rain, snow, or fog, day or night—and which objects are likely to move and which will stay put. Deep learning has helped make this task possible, at least in part, but deep learning, as always, requires a profusion of training examples.

Self-driving car companies collect these training examples from countless hours of video taken by cameras mounted on actual cars driving in traffic on highways and city streets. These cars may be self-driving prototypes being tested by companies or, in the case of Tesla, cars driven by customers who, upon purchase of a Tesla vehicle, must agree to a data-sharing policy with the company.[9]

Tesla owners aren't required to label every object on the videos taken by their cars. But someone has to. In 2017, the *Financial Times* reported that "most companies working on this technology employ hundreds or even thousands of people, often in offshore outsourcing centres in India or China, whose job it is to teach the robo-cars to recognize pedestrians, cyclists and other obstacles. The workers do this by manually marking up or 'labeling' thousands of hours of video footage, often frame by frame."[10] New companies have sprung up to offer labeling data as a service; Mighty AI, for example, offers "the labeled data you need to train your computer vision models" and promises "known, verified, and trusted annotators who specialize in autonomous driving data."[11]

The Long Tail

The supervised-learning approach, using large data sets and armies of human annotators, works well for at least some of the visual abilities needed for self-driving cars (many companies are also exploring the use of video-game-like driving-simulation programs to augment supervised training). But what about in the rest of life? Virtually everyone working in the AI field agrees that supervised learning is not a viable path to general-purpose AI.

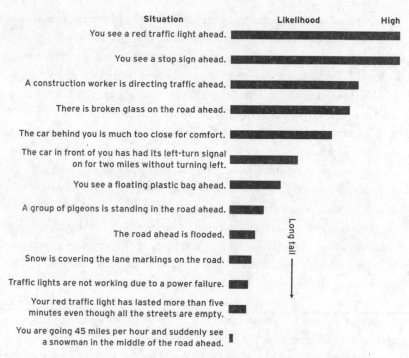

FIGURE 13: Possible situations a self-driving car might encounter, ranked by likelihood, illustrating the "long tail" of unlikely scenarios

As the renowned AI researcher Andrew Ng has warned, "Requiring so much data is a major limitation of [deep learning] today."[12] Yoshua Bengio, another high-profile AI researcher, agrees: "We can't realistically label everything in the world and meticulously explain every last detail to the computer."[13]

This issue is compounded by the so-called long-tail problem: the vast range of possible unexpected situations an AI system could be faced with. Figure 13 illustrates this phenomenon by giving the likelihood of various hypothetical situations that a self-driving car might encounter during, say, a day's worth of driving. Very common situations, such as encountering a red traffic light or a stop sign, are rated as having high likelihood; medium-likelihood situations include broken glass and wind-whipped plastic bags—not encountered every day (depending on where you drive), but not uncommon. It is less likely that your self-driving car would encounter a flooded road or lane markings obscured by snow, and even less likely that you would face a snowman in the middle of a high-speed road.

I conjured up these different scenarios and guessed at their relative likelihood; I'm sure you can come up with many more of your own. Any individual car is probably safe: after all, taken together, experimental autonomous cars have driven millions of miles and have caused a relatively small number of accidents (albeit a few high-profile fatal ones). But once self-driving cars are widespread, while each individual unlikely situation is, by definition, very unlikely, there are so many possible scenarios in the world of driving and so many cars that some self-driving car somewhere is *likely* to encounter one of them at some point.

The term *long tail* comes from statistics, in which certain probability distributions are shaped like the one in figure 13: the long list of very unlikely (but possible) situations is called the "tail" of the distribution. (The situations in the tail are sometimes called edge cases.) Most real-world domains for AI exhibit this kind of long-tail phenomenon: events in the real world are usually predictable, but there remains a long tail of low-probability, unexpected occurrences. This is a problem if we rely solely on supervised learning to provide our AI system with its knowledge of the world; the situations in the tail don't show up in the training data often enough, if at all, so the system is more likely to make errors when faced with such unexpected cases.

Here are two real-world examples. In March 2016, there was a massive snowstorm forecast in the Northeast of the United States, and reports appeared on Twitter that Tesla vehicles' Autopilot mode, which enables limited autonomous driving, was getting confused between lane markings and salt lines laid out on the highway in anticipation of the storm (figure 14). In February 2016, one of Google's prototype self-driving cars, while making a right turn, had to veer to the left to avoid sandbags on the right side of a California road, and the car's left front struck a public bus driving in the left lane. Each vehicle had expected the other to yield (perhaps the bus driver expected a human driver who would be more intimidated by the much larger bus).

Companies working on autonomous-vehicle technology are acutely aware of the long-tail problem: their teams brainstorm possible long-tail

FIGURE 14:
Salt lines on a highway, in advance of a forecasted snowstorm, were reported to be confusing Tesla's Autopilot feature.

scenarios and actively create extra training examples as well as specially coded strategies for all the unlikely scenarios they can come up with. But of course it is impossible to train or code a system for all the possible situations it might encounter.

A commonly proposed solution is for AI systems to use supervised learning on small amounts of labeled data and learn everything else via unsupervised learning. The term *unsupervised learning* refers to a broad group of methods for learning categories or actions without labeled data. Examples include methods for clustering examples based on their similarity or learning a new category via analogy to known categories. As I'll describe in a later chapter, perceiving abstract similarity and analogies is something at which humans excel, but to date there are no very successful AI methods for this kind of unsupervised learning. Yann LeCun himself acknowledges that "unsupervised learning is the dark matter of AI." In other words, for general AI, almost all learning will have to be unsupervised, but no one has yet come up with the kinds of algorithms needed to perform successful unsupervised learning.

Humans make mistakes all the time, even (or especially) in driving; any one of us might have hit that public bus, had we been the one veering around sandbags. But humans also have a fundamental competence lacking in all current AI systems: common sense. We have vast background knowledge of the world, both its physical and its social aspects. We have a good sense of how objects—both inanimate and living—are likely to behave, and we use this knowledge extensively in making decisions about how to act in any given situation. We can infer the reason behind salt lines on the road even if we have never driven in snow before. We know how to interact socially with other humans, so we can use eye contact, hand signals, and other body language to deal with broken traffic lights during a power failure. We generally know to yield the road to a large public bus, even if we technically have the right of way. I've used driving as an example here, but we humans use common sense—usually subconsciously—in every facet of life. Many people believe that until AI systems have common sense as humans do, we won't be able to trust them to be fully autonomous in complex real-world situations.

What Did My Network Learn?

A few years ago, Will Landecker, then a graduate student in my research group, trained a deep neural network to classify photographs into one of two categories: "contains an animal" and "does not contain an animal." The network was trained on photos like the ones in figure 15, and it performed very well on this task on the test set. But what did the network actually learn? By performing a careful study, Will found an unexpected answer: in part, the network learned to classify images with blurry backgrounds as "contains an animal," whether or not the image actually contained an animal.[14] The nature photos in the training and test sets obeyed an important rule of photography: focus on the subject of the photo. When the subject of the photo is an animal, the animal is the focus and the background is blurred, as in figure 15A. When the subject of the photo

FIGURE 15:
Illustration of
"animal" versus
"no animal"
classification task.
Note the blurry
background in the
image on the left.

A   B

is the background, as in figure 15B, nothing is blurred. To Will's chagrin, his network hadn't learned to recognize animals; instead, it used simpler cues—such as blurry backgrounds—that were statistically associated with animals.

This is an example of a common phenomenon seen in machine learning. The machine learns what *it* observes in the data rather than what *you* (the human) might observe. If there are statistical associations in the training data, even if irrelevant to the task at hand, the machine will happily learn those instead of what you wanted it to learn. If the machine is tested on new data with the same statistical associations, it will appear to have successfully learned to solve the task. However, the machine can fail unexpectedly, as Will's network did on images of animals without a blurry background. In machine-learning jargon, Will's network "overfitted" to its specific training set, and thus can't do a good job of applying what it learned to images that differ from those it was trained on.

In recent years, several research teams have investigated whether ConvNets trained on ImageNet and other large data sets have likewise overfitted to their training data. One group showed that if ConvNets are trained on images downloaded from the web (like those in ImageNet), they perform poorly on images that were taken by a robot moving around a house with a camera.[15] It seems that random views of household objects can look very different from photos that people put on the web. Other groups have shown that superficial changes to images, such as slightly blurring or speckling an image, changing some colors, or rotating objects in the scene,

FIGURE 16: Labels assigned to photos by Google's automated photo tagger, including the infamous "Gorillas" tag

can cause ConvNets to make significant errors even when these perturbations don't affect humans' recognition of objects.[16] This unexpected fragility of ConvNets—even those that have been said to "surpass humans at object recognition"—indicates that they are overfitting to their training data and learning something different from what we are trying to teach them.

Biased AI

The unreliability of ConvNets can result in embarrassing—and potentially damaging—errors. Google suffered a public relations nightmare in 2015 after it rolled out an automated photo-tagging feature (using a ConvNet) in its Photos app. In addition to correctly tagging images with generic descriptions such as "Airplanes," "Cars," and "Graduation," the neural network tagged a selfie featuring two African Americans as "Gorillas," as shown in figure 16. (After profuse apologies, the company's short-term solution was to remove the "Gorillas" tag from the network's list of possible categories.)

FIGURE 17:
Example of a camera
face-detection
program identifying
an Asian face as
"blinking"

Such repellent and widely mocked misclassifications are embarrassing for the companies involved, but more subtle errors due to racial or gender biases have been noted frequently in vision systems powered by deep learning. Commercial face-recognition systems, for example, tend to be more accurate on white male faces than on female or nonwhite faces.[17] Camera software for face detection is sometimes prone to missing faces with dark skin and to classifying Asian faces as "blinking" (figure 17).

Kate Crawford, a researcher at Microsoft and an activist for fairness and transparency in AI, pointed out that one widely used data set for training face-recognition systems contains faces that are 77.5 percent male and 83.5 percent white. This is not surprising, because the images were downloaded from online image searches, and photos of faces that appear online are skewed toward featuring famous or powerful people, who are predominately white and male.

Of course, these biases in AI training data reflect biases in our society, but the spread of real-world AI systems trained on biased data can magnify these biases and do real damage. Face-recognition systems, for example, are increasingly being deployed as a "secure" way to identify people in credit-card transactions, airport screening, and security cameras, and it may be only a matter of time before they are used to verify identity in voting systems, among other applications. Even small differences in accuracy between racial groups can have damaging repercussions in civil rights and access to vital services.

Such biases can be mitigated in individual data sets by having humans make sure that the photos (or other kinds of data) are balanced in their representation of, say, racial or gender groups. But this requires awareness and effort on the part of the humans curating the data. Moreover, it is often hard to tease out subtle biases and their effects. For example, one research group noted that their AI system—trained on a large data set of photos of people in different situations—would sometimes mistakenly classify a man as "woman" when the man was standing in a kitchen, an environment in which the data set had more examples of women.[18] In general, this kind of subtle bias can be apparent after the fact but hard to detect ahead of time.

The problem of bias in applications of AI has been getting a lot of attention recently, with many articles, workshops, and even academic research institutes devoted to this topic. Should the data sets being used to train AI accurately mirror our own biased society—as they often do now—or should they be tinkered with specifically to achieve social reform aims? And who should be allowed to specify the aims or do the tinkering?

Show Your Work

Remember back in school when your teacher would write "show your work" in red on your math homework? For me, showing my work was the least fun part of learning math but probably the most important, because showing how I derived my answer demonstrated that I had actually understood what I was doing, had grasped the correct abstractions, and had arrived at the answer for the right reasons. Showing my work also helped my teacher figure out why I made particular errors.

More generally, you can often trust that people know what they are doing if they can explain to you *how* they arrived at an answer or a decision. However, "showing their work" is something that deep neural networks—the bedrock of modern AI systems—cannot easily do. Let's consider the "dog" and "cat" object-recognition task I described in chapter 4. Recall that a convolutional neural network decides what object is contained in an input image by performing a sequence of mathematical operations (convolutions) propagated through many layers. For a reasonably sized network, these can amount to billions of arithmetic operations.

While it would be easy to program the computer to print out a list of all the additions and multiplications performed by a network for a given input, such a list would give us humans *zero* insight into how the network arrived at its answer. A list of a billion operations is not an explanation that a human can understand. Even the humans who train deep networks generally cannot look under the hood and provide explanations for the decisions their networks make. MIT's *Technology Review* magazine called this impenetrability "the dark secret at the heart of AI."[19] The fear is that if we don't understand how AI systems work, we can't really trust them or predict the circumstances under which they will make errors.

Humans can't always explain their thought processes either, and you generally can't look "under the hood" into other people's brains (or into their "gut feelings") to figure out how they came to any particular decision. But humans tend to trust that other humans have correctly mastered basic cognitive tasks such as object recognition and language comprehension. In part, you trust other people when you believe that their thinking is like your own. You assume, most often, that other humans you encounter have had sufficiently similar life experiences to your own, and thus you assume they are using the same basic background knowledge, beliefs, and values that you do in perceiving, describing, and making decisions about the world. In short, where other people are concerned, you have what psychologists call a theory of mind—a model of the other person's knowledge and goals in particular situations. None of us have a similar "theory of mind" for AI systems such as deep networks, which makes it harder to trust them.

It shouldn't come as a surprise then that one of the hottest new areas of AI is variously called "explainable AI," "transparent AI," or "interpretable machine learning." These terms refer to research on getting AI systems—particularly deep networks—to explain their decisions in a way that humans can understand. Researchers in this area have come up with clever ways to visualize the features that a given convolutional neural network has learned and, in some cases, to determine which parts of the input are most responsible for the output decision. Explainable AI is a field that is progressing quickly, but a deep-learning system that can successfully explain itself in human terms is still elusive.

Fooling Deep Neural Networks

There is yet another dimension to the AI trustworthiness question: Researchers have discovered that it is surprisingly easy for humans to surreptitiously trick deep neural networks into making errors. That is, if you want to deliberately fool such a system, there turn out to be an alarming number of ways to do so.

Fooling AI systems is not new. Email spammers, for example, have been in an arms race with spam-detection programs for decades. But the kinds of attacks to which deep-learning systems seem to be vulnerable are at once subtler and more troubling.

Remember AlexNet, which I discussed in chapter 5? It was the convolutional neural network that won the 2012 ImageNet challenge and that set in motion the dominance of ConvNets in much of today's AI world. If you'll recall, AlexNet's (top-5) accuracy on ImageNet was 85 percent, which blew every other competitor out of the water and shocked the computer-vision community. However, a year after AlexNet's win, a research paper appeared, authored by Christian Szegedy of Google and several others, with the deceptively mild title "Intriguing Properties of Neural Networks."[20] One of the "intriguing properties" described in the paper was that AlexNet could easily be fooled.

In particular, the paper's authors had discovered that they could take an ImageNet photo that AlexNet classified correctly with high confidence (for example, "School Bus") and distort it by making very small, specific changes to its pixels so that the distorted image looked completely unchanged to humans but was now classified with very high confidence by AlexNet as something completely different (for example, "Ostrich"). The authors called the distorted image an "adversarial example." Figure 18 shows a few samples of original images and their adversarial twins. Can't tell the difference? Congratulations! It seems that you are human.

Szegedy and his collaborators created a computer program that could, given *any* photo from ImageNet that was correctly classified by AlexNet, find specific changes to the photo to create a new adversarial example that looked unchanged to humans but caused AlexNet to give highest confidence to an incorrect category.

School bus   Ostrich   Praying mantis   Ostrich

Temple   Ostrich   Shih tzu   Ostrich

FIGURE 18: Original and "adversarial" examples for AlexNet. The left image in each pair shows the original image, which was correctly classified by AlexNet. The right image in each pair shows the adversarial example derived from this image (small changes have been made to the pixels, but the new image appears to humans to be identical to the original). Each adversarial example was confidently classified by AlexNet as "Ostrich."

Importantly, Szegedy and his collaborators found that this susceptibility to adversarial examples wasn't special to AlexNet; they showed that several other convolutional neural networks—with different architectures, hyperparameters, and training sets—had similar vulnerabilities. Calling this an "intriguing property" of neural networks is a little like calling a hole in the hull of a fancy cruise liner a "thought-provoking facet" of the ship. Intriguing, yes, and more investigation is needed, but if the leak is not fixed, this ship is going down.

Not long after the paper by Szegedy and his colleagues appeared, a group from the University of Wyoming published an article with a more direct title: "Deep Neural Networks Are Easily Fooled."[21] By using a biologically inspired computational method called genetic algorithms,[22] the Wyoming group was able to computationally "evolve" images that look like random noise to humans but for which AlexNet and other convolutional neural networks assigned specific object categories with greater than 99 percent confidence. Figure 19 shows some examples. The Wyoming group noted that deep neural networks (DNNs) "see these objects as near-perfect

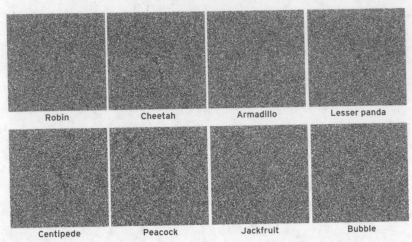

Robin   Cheetah   Armadillo   Lesser panda

Centipede   Peacock   Jackfruit   Bubble

FIGURE 19: Examples of images created by a genetic algorithm specifically to fool a convolutional neural network. In each case, AlexNet (trained on the ImageNet training set) assigned a confidence greater than 99 percent that the image was an instance of the category shown.

examples of recognizable images," which "[raises] questions about the true generalization capabilities of DNNs and the potential for costly exploits [that is, malicious applications] of solutions that use DNNs."[23]

Indeed, these two papers and subsequent related discoveries raised not only questions but also genuine alarm in the deep-learning community. If deep-learning systems, so successful at computer vision and other tasks, can easily be fooled by manipulations to which humans are not susceptible, how can we say that these networks "learn like humans" or "equal or surpass humans" in their abilities? It's clear that something very different from human perception is going on here. And if these networks are going to be used for computer vision in the real world, we'd better be darn sure that they are safeguarded from hackers using these kinds of manipulations to fool them.

All this has reenergized the small research community focusing on "adversarial learning"—that is, developing strategies that defend against potential (human) adversaries who could attack machine-learning systems. Adversarial-learning researchers often start their work by demonstrating

FIGURE 20: An AI researcher (*left*) wearing eyeglass frames with a pattern specially designed to cause a deep neural network face recognizer, trained on celebrity faces, to confidently classify the left photo as the actress Milla Jovovich (*right*). The paper describing this study gives many other examples of impersonation using "adversarial" eyeglass-frame patterns.

possible ways in which existing systems can be attacked, and some of the recent demonstrations have been stunning. In the domain of computer vision, one group of researchers developed a program that could create eyeglass frames with specific patterns that fool a face-recognition system into confidently misclassifying the wearer as another person (figure 20).[24] Another group developed small, inconspicuous stickers that could be placed on a traffic sign, resulting in a ConvNet-based vision system—similar to those used in self-driving cars—to misclassify the sign (for example, a stop sign is classified as a speed-limit sign).[25] Yet another group demonstrated a possible adversarial attack on deep neural networks for medical image analysis: they showed that it is not hard to alter an X-ray or microscopy image in a way that is imperceptible to humans but that causes a network to change its classification from, say, 99 percent confidence that the image shows no cancer to 99 percent confidence that cancer is present.[26] This group noted that such attacks could potentially be used by hospital personnel or others to create fraudulent diagnoses in order to charge insurance companies for additional (lucrative) diagnostic tests.

These are just a few examples of possible attacks that have been concocted by various research groups. Many of the possible attacks have been shown to be surprisingly robust: they work on several different networks, even when these networks are trained on different data sets. And computer vision isn't the only domain in which networks can be fooled; researchers have also designed attacks that fool deep neural networks that deal with language, including speech recognition and text analysis. We can expect that as these systems become more widely deployed in the real world, malicious users will discover many other vulnerabilities in these systems.

Understanding and defending against such potential attacks are a major area of research right now, but while researchers have found solutions for specific kinds of attacks, there is still no general defense method. Like any domain of computer security, progress so far has a "whack-a-mole" quality, where one security hole is detected and defended, but others are discovered that require new defenses. Ian Goodfellow, an AI expert who is part of the Google Brain team, says, "Almost anything bad you can think of doing to a machine-learning model can be done right now . . . and defending it is really, really hard."[27]

Beyond the immediate issue of how to defend against attacks, the existence of adversarial examples amplifies the question I asked earlier: What, precisely, are these networks learning? In particular, what are they learning that allows them to be so easily fooled? Or perhaps more important, are we fooling ourselves when we think these networks have actually learned the concepts we are trying to teach them?

To my mind, the ultimate problem is one of *understanding*. Consider figure 18, where AlexNet mistakes a school bus for an ostrich. Why would this be very unlikely to happen to a human? Even though AlexNet performs very well on ImageNet, we humans understand many things about the objects we see that are unknown to AlexNet or any other current AI system. We know what objects look like in three dimensions and can imagine this from a two-dimensional photo. We know what the function of a given object is, what role the object's parts play in its overall function, and in what contexts an object usually appears. Seeing an object brings up memories of seeing such objects in other circumstances, from other viewpoints, as well as in other sensory modalities (we remember what a given object feels like, smells like, perhaps what it sounds like when dropped, and so on). All of this background knowledge feeds into the human ability to robustly recognize a given object. Even the most suc-

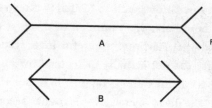

FIGURE 21: A visual illusion for humans: the horizontal line segments in *A* and *B* are the same length, but most people perceive the segment in *A* to be longer than the one in *B*.

cessful AI vision systems lack this kind of understanding and the robustness that it confers.

I've heard some AI researchers argue that humans are also susceptible to our own types of "adversarial examples": visual illusions. Like AlexNet classifying a school bus as an ostrich, humans are susceptible to perceptual errors (for example, we perceive the upper line in figure 21 to be longer than the lower line, even though both are actually the same length). But the kinds of errors that humans make are quite different from those that convolutional neural networks are susceptible to: our ability to recognize objects in everyday scenes has evolved to be very robust, because our survival depends on it. Unlike today's ConvNets, human (and animal) perception is highly regulated by *cognition*—the kind of context-dependent understanding that I described above. Moreover, ConvNets used in today's computer-vision applications are typically completely feed-forward, whereas the human visual system has many more feedback (that is, reverse direction) connections than feed-forward connections. Although neuroscientists don't yet understand the function of all this feedback, one might speculate that at least some of those feedback connections effectively prevent vulnerability to the kinds of adversarial examples that ConvNets are susceptible to. So why not just give ConvNets the same kind of feedback? This is an area of active research, but it turns out to be very difficult and hasn't produced the kind of success seen with feed-forward networks.

Jeff Clune, an AI researcher at the University of Wyoming, made a very provocative analogy when he noted that there is "a lot of interest in whether Deep Learning is 'real intelligence' or a 'Clever Hans.'"[28] Clever Hans was a horse in early twentieth-century Germany who could—his owner claimed—perform arithmetic calculations as well as understand

German. The horse responded to questions such as "What is fifteen divided by three?" by tapping his hoof the correct number of times. After Clever Hans became an international celebrity, a careful investigation eventually revealed that the horse did not actually *understand* the questions or mathematical concepts put to him, but was tapping in response to subtle, unconscious cues given by the questioner. Clever Hans has become a metaphor for any individual (or program!) that gives the appearance of understanding but is actually responding to unintentional cues given by a trainer. Does deep learning exhibit "true understanding," or is it instead a computational Clever Hans responding to superficial cues in the data? This is currently the subject of heated debates in the AI community, compounded by the fact that AI researchers don't necessarily agree on the definition of "true understanding."

On the one hand, deep neural networks, trained via supervised learning, perform remarkably well (though still far from perfectly) on many problems in computer vision, as well as in other domains such as speech recognition and language translation. Because of their impressive abilities, these networks are rapidly being taken from research settings and employed in real-world applications such as web search, self-driving cars, face recognition, virtual assistants, and recommendation systems, and it's getting hard to imagine life without these AI tools. On the other hand, it's misleading to say that deep networks "learn on their own" or that their training is "similar to human learning." Recognition of the success of these networks must be tempered with a realization that they can fail in unexpected ways because of overfitting to their training data, long-tail effects, and vulnerability to hacking. Moreover, the reasons for decisions made by deep neural networks are often hard to understand, which makes their failures hard to predict or fix. Researchers are actively working on making deep neural networks more reliable and transparent, but the question remains: Will the fact that these systems lack humanlike understanding inevitably render them fragile, unreliable, and vulnerable to attacks? And how should this factor into our decisions about applying AI systems in the real world? The next chapter explores some of the formidable challenges of balancing the benefits of AI with the risks of its unreliability and misuse.

7

.

On Trustworthy

.

and Ethical AI

.

Imagine yourself in a self-driving car, late at night, after the office Christmas party. It's dark out, and snow is falling. "Car, take me home," you say, tired and a little tipsy. You lean back, gratefully allowing your eyes to close as the car starts itself up and pulls into traffic.

All good, but how safe should you feel? The success of self-driving cars is crucially dependent on machine learning (especially deep learning), particularly for the cars' computer-vision and decision-making components. How can we determine if these cars have successfully learned all that they need to know?

This is the billion-dollar question for the self-driving car industry. I've encountered conflicting opinions from experts on how soon we can expect self-driving cars to play a significant role in daily life, with predictions ranging (at the time of this writing) from a few years to many decades. Self-driving cars have the potential to vastly improve our lives. Automated vehicles could substantially reduce the millions of annual deaths and injuries due to auto accidents, many of them caused by intoxicated or distracted drivers. In addition, automated vehicles would allow their human passengers to be productive rather than idle during commute times. These vehicles also have the potential to be more energy efficient than cars with human drivers and will be a godsend for blind or handicapped people who can't drive. But all this will come to pass only if we humans are willing to trust these vehicles with our lives.

Machine learning is being deployed to make decisions affecting the lives of humans in many domains. What assurances do you have that the machines creating your news feed, diagnosing your diseases, evaluating your loan applications, or—God forbid—recommending your prison sentence have learned enough to be trustworthy decision makers?

These are vexing questions not just for AI researchers but also for society as a whole, which must eventually weigh the many current and future positive uses of AI against concerns about its trustworthiness and misuse.

Beneficial AI

When one considers the role of AI in our society, it might be easy to focus on the downsides. However, it's essential to remember that there are huge benefits that AI systems already bring to society and that they have the potential to be even more beneficial. Current AI technology is central to services you yourself might use all the time, sometimes without even knowing that AI is involved, including speech transcription, GPS navigation and trip planning, email spam filters, language translation, credit-card fraud alerts, book and music recommendations, protection against computer viruses, and optimizing energy usage in buildings.

If you are a photographer, filmmaker, fine artist, or musician, you might be using AI systems that assist you in creative projects, such as programs that help photographers edit their photos or assist composers in music notation or arrangements. If you are a student, you might benefit from "intelligent tutoring systems" that adapt to your particular learning style. If you are a scientist, there's a good chance you have used one of the many available AI tools that help analyze your data. If you are blind or otherwise visually disabled, you might use smartphone computer-vision apps that read handwritten or printed text (for example, on signs, restaurant menus, or money). If you are hearing-impaired, you can now see quite accurate captions on YouTube videos and, in some cases, get real-time speech transcription during a lecture. These are just a few examples of the ways in which current AI tools are improving people's lives. Many additional AI technologies are still in research mode but are on the verge of becoming mainstream.

In the near future, AI applications will likely be widespread in health care. We will see AI systems assisting physicians in diagnosing diseases and in suggesting treatments; discovering new drugs; and monitoring the health and safety of the elderly in their homes. Scientific modeling and data analysis will increasingly rely on AI tools—for example, in improving models of climate change, population growth and demographic change, ecological and food science, and other major issues that society will be facing over the next century. For Demis Hassabis, the cofounder of Google's DeepMind group, this is the most important potential benefit of AI:

> We might have to come to the sobering realisation that even with the smartest set of humans on the planet working on these problems, these [problems] may be so complex that it's difficult for individual humans and scientific experts to have the time they need in their lifetimes to even innovate and advance. . . . It's my belief we're going to need some assistance and I think AI is the solution to that.[1]

We've all heard that in the future AI will take over the jobs that humans hate—low-wage jobs that are boring, exhausting, degrading, exploitative, or downright dangerous. If this actually happens, it could be a true boon for human well-being. (Later I'll discuss the other side of this coin—AI taking away *too many* human jobs.) Robots are already widely used for menial and repetitive factory tasks, though there are many such jobs still beyond the abilities of today's robots. But as AI progresses, more and more of these jobs could be taken over by automation. Examples of future AI workplace applications include self-driving trucks and taxis, as well as robots for harvesting fruits, fighting fires, removing land mines, and performing environmental cleanups. In addition, robots will likely see an even larger role than they have now in planetary and space exploration.

Will it actually benefit society for AI systems to take over such jobs? We can look to the history of technology to give us some perspective. Here are a few examples of jobs that humans used to do but that technology automated long ago, at least in developed countries: clothes washer; rickshaw driver; elevator operator; *punkawallah* (a servant in India whose sole job was to work a manual fan for cooling the room, before the days of electric fans); computer (a human, usually female, who performed tedious calculations by hand, particularly during World War II). Most people will agree that in those instances replacing humans with machines in such jobs made life better all around. One could argue that today's AI is simply extending that same arc of progress: improving life for humans by increasingly automating the necessary jobs that no one wants to do.

The Great AI Trade-Off

The AI researcher Andrew Ng has optimistically proclaimed, "AI is the new electricity." Ng explains further: "Just as electricity transformed almost everything 100 years ago, today I actually have a hard time thinking of an industry that I don't think AI will transform in the next several years."[2] This is an appealing analogy: the idea that soon AI will be as necessary—and as invisible—in our electronic devices as electricity itself. However, a major difference is that the science of electricity was well understood before it was widely commercialized. We are good at predicting the behavior of electricity. This is not the case for many of today's AI systems.

This brings us to what you might call the Great AI Trade-Off. Should we embrace the abilities of AI systems, which can improve our lives and even help save lives, and allow these systems to be employed ever more extensively? Or should we be more cautious, given current AI's unpredictable errors, susceptibility to bias, vulnerability to hacking, and lack of transparency in decision-making? To what extent should humans be required to remain in the loop in different AI applications? What should we require of an AI system in order to trust it enough to let it work autonomously? These questions are still hotly debated, even as AI is increasingly deployed and its promised future applications (for example, self-driving cars) are touted as being just over the horizon.

The lack of general agreement on these issues was underscored by a recent study carried out by the Pew Research Center.[3] In 2018, Pew analysts canvassed nearly one thousand "technology pioneers, innovators, developers, business and policy leaders, researchers and activists," asking them to reply to these questions:

By 2030, do you think it is most likely that advancing AI and related technology systems will enhance human capacities and empower them? That is, most of the time, will most people be better off than they are today? Or is it most likely that advancing AI and related technology systems will lessen human autonomy and agency to such an extent that most people will not be better off than the way things are today?

The respondents were divided: 63 percent predicted that progress in AI would leave humans better off by 2030, while 37 percent disagreed. Opinions ranged from the view that AI "can virtually eliminate global poverty, massively reduce disease and provide better education to almost everyone on the planet" to predictions of an apocalyptic future: legions of jobs taken over by automation, erosion of privacy and civil rights due to AI surveillance, amoral autonomous weapons, unchecked decisions by opaque and untrustworthy computer programs, magnification of racial and gender bias, manipulation of the mass media, increase of cybercrime, and what one respondent called "true, existential irrelevance" for humans.

Machine intelligence presents a knotty array of ethical issues, and discussions related to the ethics of AI and big data have filled several books.[4] In order to illustrate the complexity of the issues, I'll dig deeper into one example that is getting a lot of attention these days: automated face recognition.

The Ethics of Face Recognition

Face recognition is the task of labeling a face in an image or video (or real-time video stream) with a name. Facebook, for example, applies a face-recognition algorithm to every photo that is uploaded to its site, trying to detect the faces in the photo and to match them with known users (at least those users who haven't disabled this feature).[5] If you are on Facebook and someone posts a photo that includes your face, the system might ask you if you want to "tag yourself" in the photo. The accuracy of Facebook's face-recognition algorithm can be simultaneously impressive and creepy. Not surprisingly, this accuracy comes from using deep convolutional neural networks. The software can often recognize faces not only when the face is front and center in a photo but even when a person is one of many in a crowd.

Face-recognition technology has many potential upsides, including helping people search through their photo collections, enabling users with vision impairments to identify the people they encounter, locating missing children or criminal fugitives by scanning photos and videos for their faces, and detecting identity theft. However, it's just as easy to imagine applications that many people find offensive or threatening. Amazon, for example, markets its face-recognition system (with the strangely dystopian-sounding name Rekognition) to police departments, which can compare, say, security-camera footage with a database of known offenders or likely suspects.

Privacy is an obvious issue. Even if I'm not on Facebook (or any other social media platform with face recognition), photos including me might be tagged and later automatically recognized on the site, without my permission. Consider FaceFirst, a company that offers face-recognition services for a fee. As reported by the magazine *New Scientist*, "Face First . . . is rolling out a system for retailers that it says will 'boost sales by recognizing high-value customers each time they shop' and send 'alerts when known litigious individuals enter any of your locations.'"[6] Many other companies offer similar services.

Loss of privacy is not the only danger here. An even larger worry is reliability: face-recognition systems can make errors. If your face is matched in error, you might be barred from a store or an airplane flight or wrongly accused of a crime. What's more, present-day face-recognition systems have been shown to have a significantly higher error rate on people of color than on white people. The American Civil Liberties Union (ACLU), which vigorously opposes the use of face-recognition technology for law enforcement on civil rights grounds, tested Amazon's Rekognition system (using its default settings) on the 535 members of the U.S. Congress, comparing a photo of each member against a database of people who have been arrested on criminal charges. They found that the system incorrectly matched 28 out of the 535 members of Congress with people in the criminal database. Twenty-one percent of the errors were on photos of African American representatives (African Americans make up only about 9 percent of Congress).[7]

Amid the fallout from the ACLU's tests and other studies showing the unreliability and biases of face recognition, several high-tech companies have announced that they oppose using face recognition for law enforcement and surveillance. For example, Brian Brackeen, the CEO of the face-recognition company Kairos, wrote the following in a widely circulated article:

> Facial recognition technologies, used in the identification of suspects, negatively affects people of color. To deny this fact would be a lie. . . . I (and my company) have come to believe that the use of commercial facial recognition in law enforcement or in government surveillance of any kind is wrong—and that it opens the door for gross misconduct by the morally corrupt. . . . We deserve a world where we're not empowering governments to categorize, track and control citizens.[8]

In a blog post on his company's website, Microsoft's president and chief legal officer, Brad Smith, called for Congress to regulate face recognition:

> Facial recognition technology raises issues that go to the heart of fundamental human rights protections like privacy and freedom of expression.

These issues heighten responsibility for tech companies that create these products. In our view, they also call for thoughtful government regulation and for the development of norms around acceptable uses. Facial recognition will require the public and private sectors alike to step up—and to act.[9]

Google followed suit, announcing that it would not offer general-purpose face-recognition services via its cloud AI platform until the company can "ensure its use is aligned with our principles and values, and avoids abuse and harmful outcomes."[10]

The response of these companies is encouraging, but it brings to the forefront another vexing issue: To what extent should AI research and development be regulated, and who should do the regulating?

Regulating AI

Given the risks of AI technologies, many practitioners of AI, myself included, are in favor of some kind of regulation. But the regulation shouldn't be left solely in the hands of AI researchers and companies. The problems surrounding AI—trustworthiness, explainability, bias, vulnerability to attack, and morality of use—are social and political issues as much as they are technical ones. Thus, it is essential that the discussion around these issues include people with different perspectives and backgrounds. Simply leaving regulation up to AI practitioners would be as unwise as leaving it solely up to government agencies.

In one example of the complexity of crafting such regulations, in 2018 the European Parliament enacted a regulation on AI that some have called the "right to explanation."[11] This regulation requires, in the case of "automated decision making," "meaningful information about the logic involved" in any decision that affects an EU citizen. This information is required to be communicated "in a concise, transparent, intelligible and easily accessible form, using clear and plain language."[12] This opens the floodgates for interpretation. What counts as "meaningful information" or "the logic involved"? Does this regulation prohibit the use of hard-to-explain deep-learning methods in making decisions that affect individuals (such as loans and face recognition)? Such uncertainties will no doubt ensure gainful employment for policy makers and lawyers for a long time to come.

I believe that regulation of AI should be modeled on the regulation of other technologies, particularly those in biological and medical sciences, such as genetic engineering. In those fields, regulation—such as quality assurance and the analysis of risks and benefits of technologies—occurs via cooperation among government agencies, companies, nonprofit organizations, and universities. Moreover, there are now established fields of bioethics and medical ethics, which have considerable influence on decisions about the development and application of technologies. AI research and its applications very much need a well-thought-out regulatory and ethics infrastructure.

This infrastructure is just beginning to be formed. In the United States, state governments are starting to look into creating regulations, such as those for face recognition or self-driving vehicles. However, for the most part, the universities and the companies that create AI systems have been left to regulate themselves.

A number of nonprofit think tanks have cropped up to fill the void, often funded by wealthy tech entrepreneurs who are worried about AI. These organizations—with names such as Future of Humanity Institute, Future of Life Institute, and Centre for the Study of Existential Risk—hold workshops, sponsor research, and create educational materials and policy suggestions on the topics of safe and ethical uses of AI. An umbrella organization, called the Partnership on AI, has been trying to bring together such groups to "serve as an open platform for discussion and engagement about AI and its influences on people and society."[13]

One stumbling block is that there is no general agreement in the field on what the priorities for developing regulation and ethics should be. Should the immediate focus be on algorithms that can explain their reasoning? On data privacy? On robustness of AI systems to malicious attacks? On bias in AI systems? On the potential "existential risk" from superintelligent AI? My own opinion is that too much attention has been given to the risks from superintelligent AI and far too little to deep learning's lack of reliability and transparency and its vulnerability to attacks. I will say more about the idea of superintelligence in the final chapter.

Moral Machines

So far, my discussion has focused on ethical issues of how humans use AI. But there's another important question: Could machines themselves be able to have their own sense of morality, complete enough for us to allow them to make ethical decisions on their own, without humans having to oversee them? If we are going to give decision-making autonomy to face-recognition systems, self-driving cars, elder-care robots, or even robotic soldiers, don't we need to give these machines the same ability to deal with ethical and moral questions that we humans have?

People have been thinking about "machine morality" for as long as they've been thinking about AI.[14] Probably the best-known discussion of machine morality comes from Isaac Asimov's science fiction stories, in which he proposed the three "fundamental Rules of Robotics":

1. A robot may not injure a human being, or, through inaction, allow a human being to come to harm.
2. A robot must obey the orders given to it by human beings except where such orders would conflict with the First Law.
3. A robot must protect its own existence, as long as such protection does not conflict with the First or Second Law.[15]

These laws have become famous, but in truth, Asimov's purpose was to show how such a set of rules would inevitably fail. "Runaround," the 1942 story in which Asimov first introduced these laws, features a situation in which a robot, following the second law, moves toward a dangerous substance, at which point the third law kicks in, so the robot moves away, at which point the second law kicks in again, trapping the robot in an endless loop, resulting in a near disaster for the robot's human masters. Asimov's stories often focused on the unintended consequences of programming ethical rules into robots. Asimov was prescient: as we've seen, the problem of incomplete rules and unintended consequences has hamstrung all approaches to rule-based AI intelligence; moral reasoning is no different.

The science fiction writer Arthur C. Clarke used a similar plot device in his 1968 book, *2001: A Space Odyssey.*[16] The artificially intelligent computer HAL is programmed to always be truthful to humans, but at the same time to withhold the truth from human astronauts about the actual purpose of their space mission. HAL, unlike Asimov's clueless robot, suffers from the psychological pain of this cognitive dissonance: "He was . . . aware of the conflict that was slowly destroying his integrity—the conflict between truth, and concealment of truth."[17] The result is a computer "neurosis" that turns HAL into a killer. Reflecting on real-life machine morality, the mathematician Norbert Wiener noted as long ago as 1960 that "we had better be quite sure that the purpose put into the machine is the purpose which we really desire."[18]

Wiener's comment captures what is called the value alignment problem in AI: the challenge for AI programmers to ensure that their systems' values align with those of humans. But what are the values of humans? Does it even make sense to assume that there are universal values that society shares?

Welcome to Moral Philosophy 101. We'll start with every moral philosophy student's favorite thought experiment, the trolley problem: You are driving a speeding trolley down a set of tracks, and just ahead you see five workers standing together in the middle of the tracks. You step on the brakes, but you find that they don't work. Fortunately, there is a spur of tracks leading off to the right. You can steer the trolley onto the spur and avoid hitting the five workers. Unfortunately, there is a single worker standing in the middle of the spur. If you do nothing, the trolley will drive straight into the five workers and kill them all. If you steer the trolley to the right, the trolley will kill the single worker. What is the moral thing to do?

The trolley problem has been a staple of undergraduate ethics classes for the last century. Most people answer that it would be morally preferable for the driver to steer onto the spur, killing the single worker and saving the group of five. But philosophers have found that a different framing of essentially the same dilemma can lead people to the opposite answer.[19] Human reasoning about moral dilemmas turns out to be very sensitive to the way in which the dilemmas are presented.

The trolley problem has recently reemerged as part of the media's coverage of self-driving cars,[20] and the question of how an autonomous vehicle should be programmed to deal with such problems has become a central talking point in discussions on AI ethics. Many AI ethics thinkers have pointed out that the trolley problem itself, in which the driver has only two horrible options, is a highly contrived scenario that no real-world driver will ever encounter. But the trolley problem has become a kind of symbol for asking about how we should program self-driving cars to make moral decisions on their own.

In 2016, three researchers published results from surveys of several hundred people who were given trolley-problem-like scenarios that involved self-driving cars, and were asked for their views of the morality of different actions. In one survey, 76 percent of participants answered that it would be morally preferable for a self-driving car to sacrifice one passenger rather than killing ten pedestrians. But when asked if they would buy a self-driving car programmed to sacrifice its passengers in order to save a much larger number of pedestrians, the overwhelming majority of survey takers responded that they themselves would not buy such a car.[21] According to the authors, "We found that participants in six Amazon Mechanical Turk studies approved of utilitarian AVs [autonomous vehicles] (that is, AVs that sacrifice their passengers for the greater good) and would like others to buy them, but they would themselves prefer to ride in AVs that protect their passengers at all costs." In his commentary on this study, the psychologist Joshua Greene noted, "Before we can put our values into machines, we have to figure out how to make our values clear and consistent."[22] This seems to be harder than we might have thought.

Some AI ethics researchers have suggested that we give up trying to directly program moral rules for machines, and instead have machines learn moral values on their own by observing human behavior.[23] However, this self-learning approach inherits all of the problems of machine learning that I described in the previous chapter.

To my mind, progress on giving computers moral intelligence cannot be separated from progress on other kinds of intelligence: the true challenge is to create machines that can actually *understand* the situations that they confront. As Isaac Asimov's stories demonstrate, a robot can't reliably follow an order to avoid harming a human unless it can understand the concept of *harm* in different situations. Reasoning about morality requires one to recognize cause-and-effect relationships, to imagine different possible futures, to have a sense of the beliefs and goals of others, and to predict the likely outcomes of one's actions in whatever situation one finds oneself. In other words, a prerequisite to trustworthy moral reasoning is general common sense, which, as we've seen, is missing in even the best of today's AI systems.

So far in this book we've seen how deep neural networks, trained on enormous data sets, can rival the visual abilities of humans in particular tasks. We've also seen some of the weaknesses of these networks, including their reliance on massive quantities of human-labeled data and their propensity to fail in very un-humanlike ways. How can we create an AI system that truly learns on its own—one that is more trustworthy because, like humans, it can reason about its current situation and plan for the future? In the next part of the book, I'll describe how AI researchers are using games such as chess, Go, and even Atari video games as "microcosms" in order to develop machines with more humanlike learning and reasoning capabilities, and I'll assess how the resulting superhuman game-playing machines might transfer their skills to the real world.

Part III

.

Learning to Play

.

8

.

Rewards for

.

Robots

.

When the journalist Amy Sutherland was doing research for a book on exotic animal trainers, she learned that their primary method is preposterously simple: "reward behavior I like and ignore behavior I don't." And as she wrote in *The New York Times'* Modern Love column, "Eventually it hit me that the same techniques might work on that stubborn but lovable species, the American husband." Sutherland wrote about how, after years of futile nagging, sarcasm, and resentment, she used this simple method to covertly train her oblivious husband to pick up his socks, find his own car keys, show up to restaurants on time, and shave more regularly.[1]

This classic training technique, known in psychology as operant conditioning, has been used for centuries on animals and humans. Operant conditioning inspired an important machine-learning approach

FIGURE 22:
A Sony Aibo robotic
dog, about to kick a
robot soccer ball

called reinforcement learning. Reinforcement learning contrasts with the supervised-learning method I've described in previous chapters: in its purest form, reinforcement learning requires no labeled training examples. Instead, an *agent*—the learning program—performs *actions* in an *environment* (usually a computer simulation) and occasionally receives *rewards* from the environment. These intermittent rewards are the only feedback the agent uses for learning. In the case of Amy Sutherland's husband, the rewards were her smiles, kisses, and words of praise. While a computer program might not respond to a kiss or an enthusiastic "you're the greatest," it can be made to respond to a machine equivalent of such appreciation—such as positive numbers added to its memory.

While reinforcement learning has been part of the AI toolbox for decades, it has long been overshadowed by neural networks and other supervised-learning methods. This changed in 2016 when reinforcement learning played a central role in a stunning and momentous achievement in AI: a program that learned to beat the best humans at the complex game of Go. In order to explain that program, as well as other recent achievements of reinforcement learning, I'll first take you through a simple example to illustrate how reinforcement learning works.

Training Your Robo-Dog

For our illustrative example, let's look to the fun game of robot soccer, in which humans (usually college students) program robots to play a simplified version of soccer on a room-sized "field." Sometimes the players are cute doglike Aibo robots like the one shown in figure 22. An Aibo robot (made by Sony) has a camera to capture visual inputs, an internal programmable computer, and a collection of sensors and motors that enable it to walk, kick, head-butt, and even wag its plastic tail.

Imagine that we want to teach our robo-dog the simplest soccer skill: when facing the ball, walk over to it, and kick it. A traditional AI approach would be to program the robot with the following rules: Take a step toward the ball. Repeat until one of your feet is touching the ball. Then kick the ball with that foot. Of course, shorthand descriptions such as "take a step toward the ball," "until one of your feet is touching the ball," and "kick the ball" must be carefully translated into detailed sensor and motor operations built into the Aibo.

Such explicit rules might be sufficient for a task as simple as this one. However, the more "intelligent" you want your robot to be, the harder it is to manually specify rules for behavior. And of course, it's impossible to devise a set of rules that will work in every situation. What if there is a large puddle between the robot and the ball? What if a soccer cone is blocking the robot's vision? What if a rock is blocking the ball's movement? As always, the real world is awash with hard-to-predict edge cases. The promise of reinforcement learning is that the agent—here our robo-dog—can learn flexible strategies on its own simply by performing actions in the world and occasionally receiving rewards (that is, *reinforcement*) without humans having to manually write rules or directly teach the agent every possible circumstance.

Let's call our robo-dog Rosie, after my favorite television robot, the wry robotic housekeeper from the classic cartoon *The Jetsons*.[2] To make things easier for this example, let's assume that Rosie comes from the factory preprogrammed with the following ability: if a soccer ball is in Rosie's line of sight, she can estimate the number of steps she would need to take to get to the ball. This number is called the "state." In general, the state of an agent at a given time is the agent's perception of its current situation. Rosie is the simplest of possible agents, in that her state is a single number. When I say that Rosie is "in" a given state x, I mean that she is currently estimating that she is x steps away from the ball.

In addition to being able to identify her state, Rosie has three built-in *actions* she can perform: she can take a step *Forward*, take a step *Backward*, and she can *Kick*. (If Rosie happens to step out-of-bounds, she is programmed to immediately step back in.) In the spirit of operant conditioning, let's give Rosie a reward only when she succeeds in kicking the ball. Note that Rosie doesn't know ahead of time which, if any, states or actions will lead to rewards.

Given that Rosie is a robot, her "reward" is simply a number, say, 10, added to her "reward memory." We can consider the number 10 the robot equivalent of a dog treat. Or perhaps not. Unlike a real dog, Rosie has no intrinsic *desire* for treats, positive numbers, or anything else. As I'll detail below, in reinforcement learning, a human-created algorithm guides Rosie's process of learning in response to rewards; that is, the algorithm tells Rosie *how* to learn from her experiences.

Reinforcement learning occurs by having Rosie take actions over a series of learning *episodes*, each of which consists of some number of *iterations*. At each iteration, Rosie determines her current state and chooses an action to take. If Rosie receives a reward, she then *learns* something, as I'll illustrate below. Here I'll let each episode last until Rosie manages to kick the ball, at which time she receives a reward. This might take a long time. As in training a real dog, we have to be patient.

Figure 23 illustrates a hypothetical learning episode. The episode begins with the trainer (me) placing Rosie and the ball in some initial locations on the field, with Rosie facing the ball (figure 23A). Rosie determines her current state: twelve steps away from the ball. Because Rosie hasn't learned anything yet, our dog, an innocent "tabula rasa," doesn't know which action should be preferred, so she chooses an action at random from her three possibilities: *Forward*, *Backward*, *Kick*. Let's say she chooses *Backward* and takes a step back. We humans can see that *Backward* is a bad action to take, but remember, we're letting Rosie figure out on her own how to perform this task.

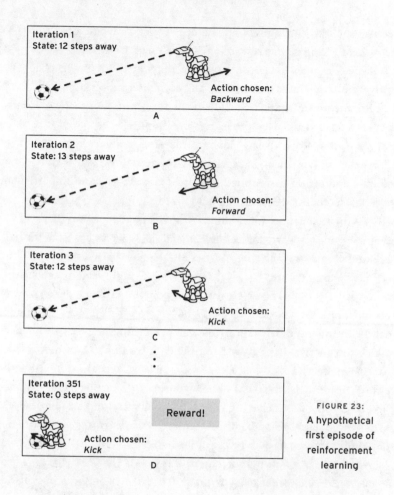

FIGURE 23:
A hypothetical
first episode of
reinforcement
learning

At iteration 2 (figure 23B), Rosie determines her new state: thirteen steps from the ball. She then chooses a new action to take, again at random: *Forward*. At iteration 3 (figure 23C), Rosie determines her "new" state: twelve steps away from the ball. She's back to square one, but Rosie doesn't even know that she has been in this state before! In the purest form of reinforcement learning, the learning agent doesn't remember its previous states. In general, remembering previous states might take a lot of memory and doesn't turn out to be necessary.

At iteration 3, Rosie—again at random—chooses the action *Kick*, but because she's kicking empty air, she doesn't get a reward. She has yet to learn that kicking gives a reward only if she's next to the ball.

Rosie continues to choose random actions, without any feedback, for many iterations. But at some point, let's say at iteration 351, just by dumb luck Rosie ends up next to the ball and chooses *Kick* (figure 23D). Finally, she gets a reward and uses it to learn something.

What does Rosie learn? Here we take the simplest approach to reinforcement learning: upon receiving a reward, Rosie learns only about the state and action that immediately preceded the reward. In particular, Rosie learns that if she is in that state (for example, zero steps from the ball), taking that action (for example, *Kick*) is a good idea. But that's all she learns. She doesn't learn, for example, that if she is zero steps from the ball, *Backward* would be a *bad* choice. After all, she hasn't tried that yet. For all she knows, taking a step backward in that state might lead to a much bigger reward! Rosie also doesn't learn at this point that if she is *one* step away, *Forward* would be a good choice. She has to wait for the next episode for that. Learning too much at one time can be detrimental; if Rosie happens to kick the air two steps away from the ball, we don't want her to learn that this ineffective kick was actually a necessary step toward getting the reward. In humans, this kind of behavior might be called superstition—namely, erroneously believing that a particular action can help cause a particular good or bad outcome. In reinforcement learning, superstition is something that you have to be careful to avoid.

A crucial notion in reinforcement learning is that of the *value of performing a particular action in a given state*. The *value* of action *A* in state *S* is a number reflecting the agent's current prediction of how much reward it will eventually obtain if, when in state *S*, it performs action *A*, and then continues performing high-value actions. Let me explain. If your current state is "holding a chocolate in your hand," an action with high value would be to bring your hand to your mouth. Subsequent actions with high value would be to open your mouth, put the chocolate inside, and chew. Your reward is the delicious sensation of eating the chocolate. Bringing your hand to your mouth doesn't immediately produce this reward, but this action is on the right path, and if you've eaten chocolate before, you can predict the intensity of the upcoming reward. The goal of reinforcement learn-

State	0 steps away		1 step away	...	10 steps away	...

Q-table:

Action	Forward 0	Forward 0	...	Forward 0	...
	Backward 0	Backward 0		Backward 0	
	Kick 10	Kick 0		Kick 0	

FIGURE 24: Rosie's Q-table after her first episode of reinforcement learning

ing is for the agent to learn values that are good predictions of upcoming rewards (assuming that the agent keeps doing the right thing after taking the action in question).[3] As we'll see, the process of learning the values of particular actions in a given state typically takes many steps of trial and error.

Rosie keeps track of the values of actions in a big table in her computer memory. This table, illustrated in figure 24, lists all the possible states for Rosie (that is, all possible distances she could be from the ball, up to the length of the field), and for each state, her possible actions. Given a state, each action in that state has a numerical value; these values will change—becoming more accurate predictions of upcoming rewards—as Rosie continues to learn. This table of states, actions, and values is called the Q-table. This form of reinforcement learning is sometimes called Q-learning. The letter Q is used because the letter V (for *value*) was used for something else in the original paper on Q-learning.[4]

At the beginning of Rosie's training, I initialize the Q-table by setting all the values to 0—a "blank slate." When Rosie receives a reward for kicking the ball at the end of episode 1, the value of the action *Kick* when in state "zero steps away" is updated to 10, the value of the reward. In the future, when Rosie is in the "zero steps away" state, she can look at the Q-table, see that *Kick* has the highest value—that is, it predicts the highest reward—and decide to choose *Kick* rather than choosing randomly. That's all that "learning" means here!

Episode 1 ended with Rosie finally kicking the ball. We now move on to episode 2 (figure 25), which starts with Rosie and the ball in new locations (figure 25A). Just as before, at each iteration Rosie determines

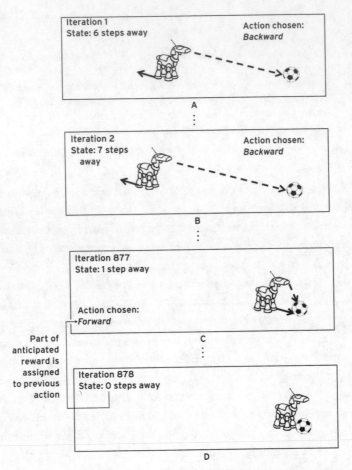

FIGURE 25: The second episode of reinforcement learning

her current state—initially, six steps away—and chooses an action, now by looking in her Q-table. But at this point, the values of actions in her current state are still all 0s; there's no information yet to help her choose among them. So Rosie again chooses an action at random: *Backward*. And she chooses *Backward* again at the next iteration (figure 25B). Our robo-dog's training has a long way to go.

Everything continues as before, until Rosie's floundering random trial-and-error actions happen to land her one step away from the ball

		Q-table:			
State	**0 steps away**	**1 step away** ...		**10 steps away**	...
Action	*Forward* 0	*Forward* 8	...	*Forward* 0	...
	Backward 0	*Backward* 0		*Backward* 0	
	Kick 10	*Kick* 0		*Kick* 0	

FIGURE 26: Rosie's Q-table after her second episode of reinforcement learning

(figure 25C), and she happens to choose *Forward*. Suddenly Rosie finds her foot next to the ball (figure 25D), and the Q-table has something to say about this state. In particular, it says that her current state—zero steps from the ball—has an action—*Kick*—that is predicted to lead to a reward of 10. Now she can use this information, learned at the previous episode, to choose an action to perform, namely *Kick*. But here's the essence of Q-learning: Rosie can now learn something about the action (*Forward*) she took in the immediately *previous* state (one step away). That is what led her to be in the excellent position she is in now! Specifically, the value of action *Forward* in the state "one step away" is updated in the Q-table to have a higher value, some fraction of the value of the action "*Kick* when zero steps away," which directly leads to a reward. Here I've updated this value to 8 (figure 26).

The Q-table now tells Rosie that it's really good to kick when in the "zero steps away" state and that it's almost as good to step forward when in the "one step away" state. The next time Rosie finds herself in the "one step away" state, she'll have some information about what action she should take, as well as the ability to learn an update for the immediately past action—the *Forward* action in the "two steps away" state. Note that it is important for these learned action values to be reduced ("discounted") as they go back in time from the actual reward; this allows the system to learn an efficient path to an actual reward.

Reinforcement learning—here, the gradual updating of values in the Q-table—continues, episode to episode, until Rosie has finally learned to perform her task from any initial starting point. The Q-learning algorithm is a way to assign values to actions in a given state, including those actions that don't lead directly to rewards but that set the stage for the relatively rare states in which the agent does receive rewards.

I wrote a program that simulated Rosie's Q-learning process as described above. At the beginning of each episode, Rosie was placed, facing the ball, a random number of steps away (with a maximum of twenty-five and a minimum of zero steps away). As I mentioned earlier, if Rosie stepped out-of-bounds, my program simply has her step back in. Each episode ended when Rosie succeeded in reaching and kicking the ball. I found that it took about three hundred episodes for her to learn to perform this task perfectly, no matter where she started.

This "training Rosie" example captures much of the essence of reinforcement learning, but I left out many issues that reinforcement-learning researchers face for more complex tasks.[5] For example, in real-world tasks, the agent's perception of its state is often uncertain, unlike Rosie's perfect knowledge of how many steps she is from the ball. A real soccer-playing robot might have only a rough estimate of distance, or even some uncertainty about which light-colored, small object on the soccer field is actually the ball. The effects of performing an action can also be uncertain: for example, a robot's *Forward* action might move it different distances depending on the terrain, or even result in the robot falling down or colliding with an unseen obstacle. How can reinforcement learning deal with uncertainties like these?

Additionally, how should the learning agent choose an action at each time step? A naive strategy would be to always choose the action with the highest value for the current state in the Q-table. But this strategy has a problem: it's possible that other, as-yet-unexplored actions will lead to a higher reward. How often should you explore—taking actions that you haven't yet tried—and how often should you choose actions that you already expect to lead to some reward? When you go to a restaurant, do you always order the meal you've already tried and found to be good, or do you try something new, because the menu might contain an even better option? Deciding how much to *explore* new actions and how much to *exploit* (that is, stick with) tried-and-true actions is called the exploration versus exploitation balance. Achieving the right balance is a core issue for making reinforcement learning successful.

These are samples of ongoing research topics among the growing community of people working on reinforcement learning. Just as in the field of deep learning, designing successful reinforcement-learning systems is still a difficult (and sometimes lucrative!) art, mastered by a relatively small group of experts who, like their deep-learning counterparts, spend a lot of time tuning hyperparameters. (How many learning episodes should be allowed? How many iterations per episode should be allowed? How much should a reward be "discounted" as it is spread back in time? And so on.)

Stumbling Blocks in the Real World

Setting these issues aside for now, let's look at two major stumbling blocks that might arise in extrapolating our "training Rosie" example to reinforcement learning in real-world tasks. First, there's the Q-table. In complex real-world tasks—think, for example, of a robot car learning to drive in a crowded city—it's impossible to define a small set of "states" that could be listed in a table. A single state for a car at a given time would be something like the entirety of the data from its cameras and other sensors. This means that a self-driving car effectively faces an infinite number of possible states. Learning via a Q-table like the one in the "Rosie" example is out of the question. For this reason, most modern approaches to reinforcement learning use a neural network instead of a Q-table. The neural network's job is to learn what values should be assigned to actions in a given state. In particular, the network is given the current state as input, and its outputs are its estimates of the values of all the possible actions the agent can take in that state. The hope is that the network can learn to group related states into general concepts (*It's safe to drive forward* or *Stop immediately to avoid hitting an obstacle*).

The second stumbling block is the difficulty, in the real world, of actually carrying out the learning process over many episodes, using a real robot. Even our "Rosie" example isn't feasible. Imagine yourself initializing a new episode—walking out on the field to set up the robot and the ball—hundreds of times, not to mention waiting around for the robot to perform its hundreds of actions per episode. You just wouldn't have enough time. Moreover, you might risk the robot damaging itself by choosing the wrong action, such as kicking a concrete wall or stepping forward over a cliff.

Just as I did for Rosie, reinforcement-learning practitioners almost always deal with this problem by building *simulations* of robots and environments and performing all the learning episodes in the simulation rather than in the real world. Sometimes this approach works well. Robots have been trained using simulations to walk, hop, grasp objects, and drive a remote-control car, among other tasks, and the robots were able, with various levels of success, to transfer the skills learned during simulation to the real world.[6] However, the more complex and unpredictable the environment, the less successful are the attempts to transfer what is learned in simulation to the real world. Because of these difficulties, it makes sense that to date the greatest successes of reinforcement learning have been not in robotics but in domains that can be perfectly simulated on a computer. In particular, the best-known reinforcement-learning successes have been in the domain of game playing. Applying reinforcement learning to games is the topic of the next chapter.

9

·

Game On

·

Since the earliest days of AI, enthusiasts have been obsessed with creating programs that can beat humans at games. In the late 1940s, both Alan Turing and Claude Shannon, two founders of the computer age, wrote programs to play chess before there were even computers that could run their code. In the decades that followed, many a young game fanatic has been driven to learn to program in order to get computers to play their favorite game, whether it be checkers, chess, backgammon, Go, poker, or, more recently, video games.

In 2010, a young British scientist and game enthusiast named Demis Hassabis, along with two close friends, launched a company in London called DeepMind Technologies. Hassabis is a colorful and storied figure in the modern AI world. A chess prodigy who was winning championships

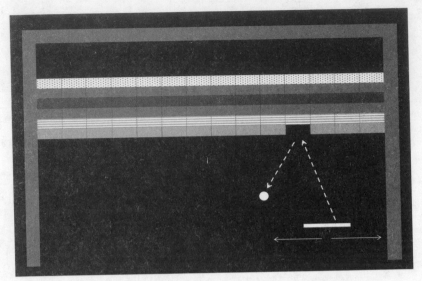

FIGURE 27: An illustration of Atari's *Breakout* game

by the age of six, he started programming video games professionally at fifteen and founded his own video game company at twenty-two. In addition to his entrepreneurial activities, he obtained a PhD in cognitive neuroscience from University College London in order to further his goal of building brain-inspired AI. Hassabis and his colleagues founded DeepMind Technologies in order to "tackle [the] really fundamental questions" about artificial intelligence.[1] Perhaps not surprisingly, the DeepMind group saw video games as the proper venue for tackling those questions. Video games are, in Hassabis's view, "like microcosms of the real world, but . . . cleaner and more constrained."[2]

Whatever your stance on video games, if you are going more for "clean and constrained" and less for "real world," you might consider creating AI programs to play Atari video games from the 1970s and '80s. This is exactly what the group at DeepMind decided to do. Depending on your age and interests, you might remember some of these classic games, such as *Asteroids*, *Space Invaders*, *Pong*, and *Ms. Pac-Man*. Are any of these ringing a bell? With their uncomplicated graphics and joystick controls, the games were easy enough for young children to learn but challenging enough to hold adults' interest.

Consider the single-player game called *Breakout*, illustrated in figure 27. The player uses the joystick to move a "paddle" (white rectangle at lower right) back and forth. A "ball" (white circle) can be bounced off the paddle to hit different-colored rectangular "bricks." The ball can also bounce off the gray "walls" at the sides. If the ball hits one of the bricks (patterned rectangles), the brick disappears, the player gains points, and the ball bounces back. Bricks in higher layers are worth more points than those in lower layers. If the ball hits the "ground" (bottom of the screen), the player loses one of five "lives," and if any "lives" remain, a new ball shoots into play. The player's goal is to maximize the score over the five lives.

There's an interesting side note here. *Breakout* was the result of Atari's effort to create a single-player version of its successful game *Pong*. The design and implementation of *Breakout* were originally assigned in 1975 to a twenty-year-old employee named Steve Jobs. Yes, that Steve Jobs (later, cofounder of Apple). Jobs lacked sufficient engineering skills to do a good job on *Breakout*, so he enlisted his friend Steve Wozniak, aged twenty-five (later, the other cofounder of Apple), to help on the project. Wozniak and Jobs completed the hardware design of *Breakout* in four nights, starting work each night after Wozniak had completed his day job at Hewlett-Packard. Once released, *Breakout*, like *Pong*, was hugely popular among gamers.

If you're getting nostalgic but neglected to hang on to your old Atari 2600 game console, you can still find many websites offering *Breakout* and other games. In 2013, a group of Canadian AI researchers released a software platform called the Arcade Learning Environment that made it easy to test machine-learning systems on forty-nine of these games.[3] This was the platform used by the DeepMind group in their work on reinforcement learning.

Deep Q-Learning

The DeepMind group combined reinforcement learning—in particular Q-learning—with deep neural networks to create a system that could learn to play Atari video games. The group called their approach deep

Input: Current state (current frame plus three prior frames)

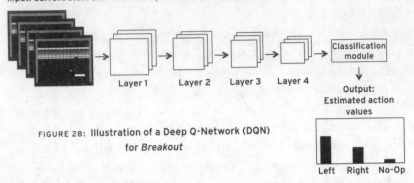

FIGURE 28: Illustration of a Deep Q-Network (DQN)
for *Breakout*

Q-learning. To explain how deep Q-learning works, I'll use *Breakout* as a running example, but DeepMind used the same method on all the Atari games they tackled. Things will get a bit technical here, so fasten your seat belt (or skip to the next section).

Recall how we used Q-learning to train Rosie the robo-dog. In an episode of Q-learning, at each iteration the learning agent (Rosie) does the following: it figures out its current state, looks up that state in the Q-table, uses the values in the table to choose an action, performs that action, possibly receives a reward, and—the learning step—updates the values in its Q-table.

DeepMind's deep Q-learning is exactly the same, except that a convolutional neural network takes the place of the Q-table. Following DeepMind, I'll call this network the Deep Q-Network (DQN). Figure 28 illustrates a DQN that is similar to (though simpler than) the one used by DeepMind for learning to play *Breakout*. The input to the DQN is the state of the system at a given time, which here is defined to be the current "frame"—the pixels of the current screen—plus three prior frames (screen pixels from three previous time steps). This definition of state provides the system with a small amount of memory, which turns out to be useful here. The outputs of the network are the estimated values for each possible action, given the input state. The possible actions are the following: move the paddle *Left*, move the paddle *Right*, and *No-Op* ("no operation," that is, don't move the paddle). The network itself is a ConvNet virtually identical to the one I described in chapter 4. Instead of the values in a Q-table, as we saw in the "Rosie" example, in deep Q-learning it is the *weights* in this network that are learned.

DeepMind's system learns to play *Breakout* over many episodes. Each episode corresponds to a play of the game, and each iteration during an episode corresponds to the system performing a single action. In particular, at each iteration the system inputs its state to the DQN and chooses an action based on the DQN's output values. The system doesn't always choose the action with the highest estimated value; as I mentioned above, reinforcement learning requires a balance between exploration and exploitation.[4] The system performs its chosen action (for example, moving the paddle some amount to the left) and possibly receives a reward if the ball happens to hit one of the bricks. The system then performs a step of learning—that is, updating the weights in the DQN via back-propagation.

How are the weights updated? This is the crux of the difference between supervised learning and reinforcement learning. As you'll recall from earlier chapters, back-propagation works by changing a neural network's weights so as to reduce the *error* in the network's outputs. With supervised learning, measuring this error is straightforward. Remember our hypothetical ConvNet back in chapter 4 whose goal was to learn to classify photos as "dog" or "cat"? If an input training photo pictured a dog but the "dog" output confidence was only 20 percent, then the error for that output would be $100\% - 20\% = 80\%$; that is, ideally, the output should have been 80 points higher. The network could calculate the error because it had a *label* provided by a human.

However, in reinforcement learning we have no labels. A given frame from the game doesn't come labeled with the action that should be taken. How then do we assign an error to an output in this case?

Here's the answer. Recall that if you are the learning agent, the *value* of an action in the current state is your estimate of how much reward you will receive by the end of the episode, if you choose this action (and continue choosing high-value actions). This estimate should be better the closer you get to the end of the episode, when you can tally up the actual rewards you received! The trick is to assume that the network's outputs at the *current* iteration are closer to being correct than its outputs at the *pre-*

vious iteration. Then *learning* consists in adjusting the network weights (via back-propagation) so as to minimize the difference between the current and the previous iteration's outputs. Richard Sutton, one of the originators of this method, calls this "learning a guess from a guess."[5] I'll amend that to "learning a guess from a *better* guess."

In short, instead of learning to match its outputs to human-given labels, the network learns to make its outputs consistent from one iteration to the next, assuming that later iterations give better estimates of value than earlier iterations. This learning method is called temporal difference learning.

To recap, here's how deep Q-learning works for the game of *Breakout* (and all the other Atari games). The system gives its current state as input to the Deep Q-Network. The Deep Q-Network outputs a value for each possible action. The system chooses and performs an action, resulting in a new state. Now the learning step takes place: the system inputs its new state to the network, which outputs a new set of values for each action. The difference between the new set of values and the previous set of values is considered the "error" of the network; this error is used by back-propagation to change the weights of the network. These steps are repeated over many episodes (plays of the game). Just to be clear, everything here—the Deep Q-Network, the virtual "joystick," and the game itself—is software running in a computer.

This is essentially the algorithm developed by DeepMind's researchers, although they used some tricks to improve it and speed it up.[6] At first, before much learning has happened, the network's outputs are quite random, and the system's game playing looks quite random as well. But gradually, as the network learns weights that improve its outputs, the system's playing ability improves, in many cases quite dramatically.

The $650 Million Agent

The DeepMind group applied their deep Q-learning method to the forty-nine different Atari games in the Arcade Learning Environment. While DeepMind's programmers used the same network architecture and hyperparameter settings for each game, their system learned each game

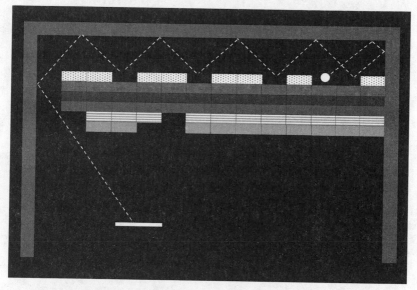

FIGURE 29: DeepMind's *Breakout* player discovered the strategy of tunneling through the bricks, which allowed it to quickly destroy high-value top bricks by bouncing off the "ceiling."

from scratch; that is, the system's knowledge (the network weights) learned for one game was not transferred when the system started learning to play the next game. Each game required training for thousands of episodes, but this could be done relatively quickly on the company's advanced computer hardware.

After a Deep Q-Network for each game was trained, DeepMind compared the machine's level of play with that of a human "professional games tester," who was allowed two hours of practice playing each game before being evaluated. Sound like a fun job? Only if you like being humiliated by a computer! DeepMind's deep Q-learning programs turned out to be better players than the human tester on more than half the games. And on half of *those* games, the programs were more than twice as good as the human. And on half of *those* games, the programs were more than five times better. One stunning example was on *Breakout*, where the DQN program scored on average more than *ten* times the human's average score.

What, exactly, did these superhuman programs learn to do? Upon investigation, DeepMind found that their programs had discovered some very clever strategies. For example, the trained *Breakout* program had discovered a devious trick, illustrated in figure 29. The program learned that if the ball was able to knock out bricks so as to build a narrow tunnel through the edge of the brick layer, then the ball would bounce back and forth between the "ceiling" and the top of the brick layer, knocking out high-value top bricks very quickly without the player having to move the paddle at all.

DeepMind first presented this work in 2013 at an international machine-learning conference.[7] The audience was dazzled. Less than a year later, Google announced that it was acquiring DeepMind for £440 million (about $650 million at the time), presumably because of these results. Yes, reinforcement learning occasionally leads to big rewards.

With a lot of money in their pockets and the resources of Google behind them, DeepMind—now called Google DeepMind—took on a bigger challenge, one that had in fact long been considered one of AI's "grand challenges": creating a program that learns to play the game Go better than any human. DeepMind's program AlphaGo builds on a long history of AI in board games. Let's start with a brief survey of that history, which will help in explaining how AlphaGo works and why it is so significant.

Checkers and Chess

In 1949, the engineer Arthur Samuel joined IBM's laboratory in Poughkeepsie, New York, and immediately set about programming an early version of IBM's 701 computer to play checkers. If you yourself have any computer programming experience, you will appreciate the challenge he faced: as noted by one historian, "Samuel was the first person to do any serious programming on the 701 and as such had no system utilities [that is, essentially no operating system!] to call on. In particular he had no assembler and had to write everything using the op codes and addresses."[8] To translate for my nonprogrammer readers, this is something like building a house using only a handsaw and a hammer. Samuel's checkers-playing program was among the earliest machine-learning programs; indeed, it was Samuel who coined the term *machine learning*.

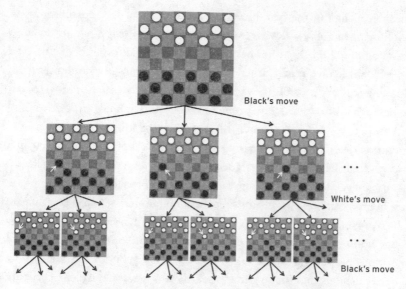

FIGURE 30: Part of a game tree for checkers. For simplicity, this figure shows only three possible moves from each board position. The white arrows point from a moved piece's previous square to its current square.

Samuel's checkers player was based on the method of searching a game tree, which is the basis of all programs for playing board games to this day (including AlphaGo, which I'll describe below). Figure 30 illustrates part of a game tree for checkers. The "root" of the tree (by convention drawn at the top, unlike the root of a natural tree) shows the initial checkerboard, before either player has moved. The "branches" from the root lead to all possible moves for the first player (here, Black). There are seven possible moves (for simplicity, the figure shows only three of these). For each of those seven moves for Black, there are seven possible response moves for White (not all shown in the figure), and so on. Each of the boards in figure 30, showing a possible arrangement of pieces, is called a board position.

Imagine yourself playing a game of checkers. At each turn, you might construct a small part of this tree in your mind. You might say to yourself, "If I make *this* move, then my opponent could make that move, in which case I could make that move, which will set me up for a jump." Most people, including the best players, consider only a few possible moves, looking ahead only a few steps before choosing which move to make. A fast computer, on the other hand, has the potential to perform this kind of look-ahead on a much larger scale. What's stopping the computer from looking at every possible move and seeing which sequence of moves most quickly leads to a win? The problem is the same kind of exponential increase we saw back in chapter 3 (remember the king, the sage, and the grains of rice?). The average game of checkers has about fifty moves, which means that the game tree in figure 30 might extend down for fifty levels. At each level, there are on average six or seven branches from each possible board position. This means that the total number of board positions in the tree could be more than six raised to the fiftieth power—a ridiculously huge number. A hypothetical computer that could look at a trillion board positions per second would take more than 10^{19} years to consider all the board positions in a single game tree. (As is often done, we can compare this number with the age of the universe, which is merely on the order of 10^{10} years.) Clearly a complete search of the game tree is not feasible.

Fortunately, it's possible for computers to play well without doing this kind of exhaustive search. On each of its turns, Samuel's checkers-playing program created (in the computer's memory) a small part of a game tree like the one in figure 30. The root of the tree was the player's current board position, and the program, using its built-in knowledge of the rules of checkers, generated all the legal moves it could make from this current board position. It then generated all the legal moves that the opponent could make from each of the resulting positions, and so on, up to four or five turns (or "plies") of look-ahead.[9]

The program then evaluated board positions that appeared at the end of the look-ahead process; in figure 30, these would be the board positions in the bottom row in the partial tree. *Evaluating* a board position means assigning it a numerical value that estimates how likely it is to lead to a win for the program. Samuel's program used an *evaluation function* that gave points for various features of the board, such as Black's advantage in total number of pieces, Black's number of kings, and how many of Black's pieces were close to being kinged. These specific features were chosen by Samuel using his knowledge of checkers. Once each of the bottom-row board positions was thus evaluated, the program employed a classic algorithm, called minimax, which used these values—from the end of the look-ahead process—in order to rate the program's immediate possible moves from its current board position. The program then chose the highest-rated move.

The intuition here is that the evaluation function will be more accurate when applied to board positions further along in the game; thus the program's strategy is to first look at all possible move sequences a few steps into the future and then apply the evaluation function to the resulting board positions. The evaluations are then propagated back up the tree by minimax, which produces a rating of all the possible immediate moves from the current board position.[10]

What the program learned was *which* features of the board should be included in the evaluation function at a given turn, as well as how to weight these different features when summing their points. Samuel experimented with several methods for learning in his system. In the most interesting version, the system learned while playing itself! The method for learning was somewhat complicated, and I won't detail it here, but it had some aspects that foreshadowed modern reinforcement learning.[11]

In the end, Samuel's checkers player impressively rose to the level of a "better-than-average player," though by no means a champion. It was characterized by some amateur players as "tricky but beatable."[12] But notably, the program was a publicity windfall for IBM: the day after Samuel demonstrated it on national television in 1956, IBM's stock price rose by fifteen points. This was the first of several times IBM saw its stock price increase after a demonstration of a game-playing program beating humans; as a more recent example, IBM's stock price similarly rose after the widely viewed TV broadcasts in which its Watson program won in the game show *Jeopardy!*

While Samuel's checkers player was an important milestone in AI history, I made this historical digression primarily to introduce three all-important concepts that it illustrates: the game tree, the evaluation function, and learning by self-play.

Deep Blue

Although Samuel's "tricky but beatable" checkers program was remarkable, especially for its time, it hardly challenged people's idea of themselves as uniquely intelligent. Even if a machine could win against human checkers champions (as one finally did in 1994[13]), mastering the game of checkers was never seen as a proxy for general intelligence. Chess is a different story. In the words of DeepMind's Demis Hassabis, "For decades, leading computer scientists believed that, given the traditional status of chess as an exemplary demonstration of human intellect, a competent computer chess player would soon also surpass all other human abilities."[14] Many people, including the early pioneers of AI Allen Newell and Herbert Simon, professed this exalted view of chess; in 1958 Newell and Simon wrote, "If one could devise a successful chess machine, one would seem to have penetrated to the core of human intellectual endeavor."[15]

Chess is significantly more complex than checkers. For example, I said above that in checkers there are, on average, six or seven possible moves from any given board position. In contrast, chess has on average thirty-five moves from any given board position. This makes the chess game tree enormously larger than that of checkers. Over the decades, chess-playing programs kept improving, in lockstep with improvements in the speed of computer hardware. In 1997, IBM had its second big game-playing triumph with Deep Blue, a chess-playing program that beat the world champion Garry Kasparov in a widely broadcast multigame match.

Deep Blue used much the same method as Samuel's checkers player: at a given turn, it created a partial game tree using the current board position as the root; it applied its evaluation function to the furthest layer in the tree and then used the minimax algorithm to propagate the values up the tree in order to determine which move it should make. The major differences between Samuel's program and Deep Blue were Deep Blue's deeper look-ahead in its game tree, its more complex (chess-specific) evaluation function, hand-programmed chess knowledge, and specialized parallel hardware to make it run very fast. Furthermore, unlike Samuel's checkers-playing program, Deep Blue did not use machine learning in any central way.

Like Samuel's checkers player before it, Deep Blue's defeat of Kasparov spurred a significant increase in IBM's stock price.[16] This defeat also generated considerable consternation in the media about the implications for superhuman intelligence as well as doubts about whether humans would still be motivated to play chess. But in the decades since Deep Blue, humanity has adapted. As Claude Shannon wrote presciently in 1950, a machine that can surpass humans at chess "will force us either to admit the possibility of mechanized thinking or to further restrict our concept of thinking."[17] The latter happened. Superhuman chess playing is now seen as something that doesn't require general intelligence. Deep Blue isn't intelligent in any sense we mean today. It can't do anything but play chess, and it doesn't have any conception of what "playing a game" or "winning" means to humans. (I once heard a speaker say, "Deep Blue may have beat Kasparov, but it didn't get any joy out of it.") Moreover, chess has survived—even prospered—as a challenging human activity. Nowadays, computer-chess programs are used by human players as a kind of training aid, in the way a baseball player might practice using a pitching machine. Is this a result of our evolving notion of intelligence, which advances in AI help to clarify? Or is it another example of John McCarthy's maxim: "As soon as it works, no one calls it AI anymore"?[18]

The Grand Challenge of Go

The game of Go has been around for more than two thousand years and is considered among the most difficult of all board games. If you're not a Go player, don't worry; none of my discussion here will require any prior knowledge of the game. But it's useful to know that the game has serious status, especially in East Asia, where it is extremely popular. "Go is a pastime beloved by emperors and generals, intellectuals and child prodigies," writes the scholar and journalist Alan Levinovitz, who goes on to quote the South Korean Go champion Lee Sedol: "There is chess in the western world, but Go is incomparably more subtle and intellectual."[19]

Go is a game that has fairly simple rules but produces what you might call emergent complexity. At each turn, a player places a piece of his or her color (black or white) on a nineteen-by-nineteen-square board, following rules for where pieces may be placed and how to capture one's opponent's pieces. Unlike chess, with its hierarchy of pawns, bishops, queens, and so on, pieces in Go ("stones") are all equal. It's the configuration of stones on the board that a player must quickly analyze to decide on a move.

Creating a program to play Go well has been a focus of AI since the field's early days. However, Go's complexity made this task remarkably hard. In 1997, the same year Deep Blue beat Kasparov, the best Go programs could still be easily defeated by average players. Deep Blue, you'll recall, was able to do a significant amount of look-ahead from any board position and then use its evaluation function to assign values to future board positions, where each value predicted whether a particular board position would lead to a win. Go programs are not able to use this strategy for two reasons. First, the size of a look-ahead tree in Go is dramatically larger than that in chess. Whereas a chess player must choose from on average 35 possible moves from a given board position, a Go player has on average 250 such possibilities. Even with special-purpose hardware, a Deep Blue–style brute-force search of the Go game tree is just not feasible. Second, no one has succeeded in creating a good evaluation function for Go board positions. That is, no one has been able to construct a successful formula that examines a board position in Go and predicts who is going to win. The best (human) Go players rely on their pattern-recognition skills and their ineffable intuition.

AI researchers haven't yet figured out how to encode intuition into an evaluation function. This is why, in 1997, the same year that Deep Blue beat Kasparov, the journalist George Johnson wrote in *The New York Times*, "When or if a computer defeats a human Go champion, it will be a sign that artificial intelligence is truly beginning to become as good as the real thing."[20] This may sound familiar—just like what people used to say about chess! Johnson quoted one Go enthusiast's prediction: "It may be a hundred years before a computer beats humans at Go—maybe even longer." A mere twenty years later, AlphaGo, which learned to play Go via deep Q-learning, beat Lee Sedol in a five-game match.

AlphaGo Versus Lee Sedol

Before I explain how AlphaGo works, let's first commemorate its spectacular wins against Lee Sedol, one of the world's best Go players. Even after watching AlphaGo defeat the then European Go champion Fan Hui half a year earlier, Lee remained confident that he would prevail: "I think [AlphaGo's] level doesn't match mine. . . . Of course, there would have been many updates in the last four or five months, but that isn't enough time to challenge me."[21]

Perhaps you were one of the more than two hundred million people who watched some part of the AlphaGo-Lee match online in March 2016. I'm certain that this ranks as the largest audience by far for any Go match in the game's twenty-five-hundred-year history. After the first game, you might have shared Lee's reaction at his loss to the program: "I am in shock, I admit that. . . . I didn't think AlphaGo would play the game in such a perfect manner."[22]

AlphaGo's "perfect" play included many moves that evoked surprise and admiration among the match's human commentators. But partway through game 2, AlphaGo made a single move that gobsmacked even the most advanced Go experts. As *Wired* reported,

> At first, Fan Hui [the aforementioned European Go champion] thought the move was rather odd. But then he saw its beauty. "It's not a human move. I've never seen a human play this move," he says. "So beautiful." It's a word he keeps repeating. Beautiful. Beautiful. Beautiful. . . . "That's a very surprising move," said one of the match's English language commentators, who is himself a very talented Go player. Then the other chuckled and said: "I thought it was a mistake." But perhaps no one was more surprised than Lee Sedol, who stood up and left the match room. "He had to go wash his face or something, just to recover," said the first commentator.[23]

Of this same move, *The Economist* noted, "Intriguingly, moves like these are sometimes made by human Go masters. They are known in Japanese as *kami no itte* ('the hand of God,' or 'divine moves')."[24]

AlphaGo won that game, and the next. But in game 4, Lee had his own *kami no itte* moment, one that captures the intricacy of the game and the intuitive power of the top players. Lee's move took the commentators by surprise, but they immediately recognized it as potentially lethal for Lee's opponent. One writer noted, "AlphaGo, however, didn't seem to realize what was happening. This wasn't something it had encountered . . . in the millions and millions of games it had played with itself. At the post-match press conference Sedol was asked what he had been thinking when he played it. It was, he said, the only move he had been able to see."[25]

AlphaGo lost game 4 but came back to win game 5 and thus the match. In the popular media, it was Deep Blue versus Kasparov all over again, with an endless supply of think pieces on what AlphaGo's triumph meant for the future of humanity. But this was even more significant than Deep Blue's win: AI had surmounted an even greater challenge than chess and had done so in a much more impressive fashion. Unlike Deep Blue, AlphaGo acquired its abilities by reinforcement learning via self-play.

Demis Hassabis noted that "the thing that separates out top Go players [is] their intuition" and that "what we've done with AlphaGo is to introduce with neural networks this aspect of intuition, if you want to call it that."[26]

How AlphaGo Works

There have been several different versions of AlphaGo, so to keep them straight, DeepMind started naming them after the human Go champions the programs had defeated—AlphaGo Fan and AlphaGo Lee—which to me evoked the image of the skulls of vanquished enemies in the collection of a digital Viking. Not what DeepMind intended, I'm sure. In any case, AlphaGo Fan and AlphaGo Lee both used an intricate mix of deep Q-learning, "Monte Carlo tree search," supervised learning, and specialized Go knowledge. But a year after the Lee Sedol match, DeepMind developed a version of the program that was both simpler than and superior to the previous versions. This newer version is called AlphaGo Zero because, unlike its predecessor, it started off with "zero" knowledge of Go besides the rules.[27] In a hundred games of AlphaGo Lee versus AlphaGo Zero, the lat-

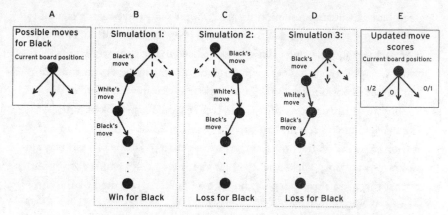

FIGURE 31: An illustration of Monte Carlo tree search

ter won every single game. Moreover, DeepMind applied the same methods (though with different networks and different built-in game rules) to learn to play both chess and *shogi* (also known as Japanese chess).[28] The authors called the collection of these methods AlphaZero. In this section, I'll describe how AlphaGo Zero worked, but for conciseness I'll simply refer to this version as AlphaGo.

The word *intuition* has an aura of mystery, but AlphaGo's intuition (if you want to call it that) arises from its combination of deep Q-learning with a clever method called "Monte Carlo tree search." Let's take a moment to unpack that cumbersome name. First, the "Monte Carlo" part. Monte Carlo is, of course, the most glamorous part of the tiny Principality of Monaco, on the French Riviera, famous for its jet-setter casinos, car racing, and frequent appearance in James Bond movies. But in science and mathematics, "Monte Carlo" refers to a family of computer algorithms, the so-called Monte Carlo method, which was first used during the Manhattan Project to help design the atomic bomb. The name comes from the idea that a degree of *randomness*—like that of the iconic spinning roulette wheel in the Monte Carlo Casino—can be used by a computer to solve difficult mathematical problems.

Monte Carlo tree search is a version of the Monte Carlo method specifically devised for computer game-playing programs. Similar to the way

Deep Blue's evaluation function worked, Monte Carlo tree search is used to assign a score to each possible move from a given board position. However, as I explained above, using extensive look-ahead in the game tree is not feasible for Go, and no one has been able to come up with a good evaluation function for board positions in Go. Monte Carlo tree search works differently.

Figure 31 illustrates Monte Carlo tree search. First, look at figure 31A. The black circle represents the current board position—that is, the configuration of pieces on the board at the current turn. Assume our Go-playing program is playing Black, and it is Black's move. Let's assume for simplicity that there are three possible moves for Black, represented by the three arrows. Which move should Black choose?

If Black had enough time, it could do a "full search" of the game tree: look ahead at *all* the possible sequences of moves that could be played and choose a move that gives the best chance of leading to a win for Black. But doing this exhaustive look-ahead isn't possible; as I mentioned earlier, even all the time since the beginning of the universe isn't nearly enough to do a full tree search in Go. With Monte Carlo tree search, Black looks ahead at only a minuscule fraction of the possible sequences that could arise from each move, counts how many wins and losses those hypothetical sequences lead to, and uses those counts to give a score to each of its possible moves. The roulette-wheel-inspired randomness is used in deciding how to do the look-ahead.

More specifically, in order to choose a move from its current position, Black "imagines" (that is, simulates) several possible ways the game could play out, as illustrated in figure 31B–D. In each of these simulations, Black starts at its current position, randomly chooses one of its possible moves, then (from the new board position) randomly chooses a move for its opponent (White), and so on, continuing until the simulated game ends up in a win or loss for Black. Such a simulation, starting from a given board position, is called a roll-out from that position.

In the figure, you can see that in the three roll-outs, Black won once and lost twice. Black can now assign a score to each possible move from its current board position (figure 31E). Move 1 (leftmost arrow) participated in two roll-outs, one of which ended in a win, so that move's score is 1 out of 2. Move 3 (rightmost arrow) participated in one roll-out, which ended in a loss, so its score is 0 out of 1. The center move was not tried at all, so its score is set to 0. Moreover, the program keeps similar statistics on all the intermediate moves that participated in the roll-outs. Once this round of Monte Carlo tree search has finished, the program can use its updated scores to decide which of its possible moves seems the most promising—here, move 1. The program can then make that move in the actual game.

When I said before that during a roll-out the program chooses moves for itself and its opponents at random, what actually happens is that the program chooses moves *probabilistically* based on any scores that those moves might have from previous rounds of Monte Carlo tree search. When each roll-out finishes with a win or loss, the algorithm updates all the scores of moves it made during that game to reflect the win or loss.

At first, the program's choice of moves from a given board position is quite random (the program is doing the equivalent of spinning a roulette wheel to choose a move), but as the program performs additional roll-outs, generating additional statistics, it is increasingly biased to choose those moves that in past roll-outs led to the most wins.

In this way, Monte Carlo tree search doesn't have to guess from just looking at a board position which move is most likely to lead to a win; it uses its roll-outs to collect statistics on how many times a given move *actually* leads to a win or loss. The more roll-outs the algorithm runs, the better its statistics. As before, the program needs to balance exploitation (choosing the highest-scoring moves during a roll-out) with exploration (sometimes choosing lower-scoring moves for which the program doesn't yet have much statistics). In figure 31, I showed three roll-outs; AlphaGo's Monte Carlo tree search performed close to two thousand roll-outs per turn.

The computer scientists at DeepMind didn't invent Monte Carlo tree search. It was first proposed in the context of game trees in 2006, and it resulted in a very big improvement in the ability of computer Go programs. But these programs still couldn't beat the best humans. One problem was that generating sufficient statistics from roll-outs can take a lot of time, especially in Go, with its vast number of possible moves. The DeepMind group realized that they could improve their system by complementing Monte Carlo tree search with a deep convolutional neural network. Given the current board position as input, AlphaGo uses a trained deep convolutional neural network to assign a rough value to all possible moves from the current position. Then Monte Carlo tree search uses those values to kick-start its search: rather than initially choosing moves at random, Monte Carlo tree search uses values output by the ConvNet as an indicator of which initial moves should be preferred. Imagine that you are AlphaGo staring at a board position: before you start the Monte Carlo process of performing roll-outs from that position, the ConvNet is whispering in your ear which of the possible moves from your current position are probably the best ones.

Conversely, the results of Monte Carlo tree search feed back to train the ConvNet. Imagine yourself as AlphaGo after a Monte Carlo tree search. The results of your search are new probabilities assigned to all your possible moves, based on how many times those moves led to wins or losses during the roll-outs you performed. These new probabilities are now used to correct your ConvNet's output, via back-propagation. You and your opponent then choose moves, as a result of which you have a new board position, and the process continues. In principle, the convolutional neural network will learn to recognize patterns, just as Go masters do. Eventually, the ConvNet will play the role of the program's "intuition," which is further improved by Monte Carlo tree search.

Like its ancestor, Samuel's checkers player, AlphaGo learns by playing against *itself* over many games (about five million). During its training, the convolutional neural network's weights are updated after each move based on the difference between the network's output values and the improved values after Monte Carlo tree search is run. Then, when it's time for AlphaGo to play, say, a human like Lee Sedol, the trained ConvNet is used at each turn to generate values to help Monte Carlo tree search get started.

With its AlphaGo project, DeepMind demonstrated that one of AI's longtime grand challenges could be conquered by an inventive combination of reinforcement learning, convolutional neural networks, and Monte Carlo tree search (and adding powerful modern computing hardware to the mix). As a result, AlphaGo has attained a well-deserved place in the AI pantheon. But what's next? Will this potent combination of methods generalize beyond the world of game playing? This is the question I discuss in the next chapter.

10

.

Beyond Games

.

Over the past decade, reinforcement learning has transformed from a relatively obscure branch of AI to one of the field's most exciting (and heavily funded) approaches. The resurgence of reinforcement learning, especially in the public eye, is largely due to the DeepMind projects I described in the previous chapter. DeepMind's results on Atari games and on Go are indeed remarkable and important, and they deserve their accolades.

However, developing superhuman game-playing programs is, for most AI researchers, not an end in and of itself. Let's step back and ask about the implications of these successes for broader progress in AI. Demis Hassabis has something to say about this:

Games are just our development platform. . . . It's the fastest way to develop these AI algorithms and test them, but ultimately we want to use them so they apply to real-world problems and have a huge impact on things like healthcare and science. The whole point is that it's general AI—it's learning how to do things [based on] its own experience and its own data.[1]

Let's dig into this a bit. How *general* is this AI, really? How applicable to the real world, beyond games? To what extent are these systems actually learning "on their own"? And what is it, exactly, that they learn?

Generality and "Transfer Learning"

When I was searching online for articles about AlphaGo, the web offered me this catchy headline: "DeepMind's AlphaGo Mastered Chess in Its Spare Time."[2] This claim is wrong and misleading, and it's important to understand why. AlphaGo (in all its versions) can't play anything but Go. Even the most general version, AlphaZero, is not a single system that learned to play Go, chess, and *shogi*. Each game has its own separate convolutional neural network that must be trained from scratch for its particular game. Unlike humans, none of these programs can "transfer" anything it has learned about one game to help it learn a different game.

The same is true for the various Atari game-playing programs: each learns its own network weights from scratch. It's as if you learned to play *Pong*, but then in order to learn to play *Breakout*, you'd have to completely forget everything you learned about playing *Pong* and start from square one.

A hopeful phrase in the machine-learning community is "transfer learning," which refers to the ability of a program to *transfer* what it has learned about one task to help it perform a different, related task. For humans, transfer learning is automatic. After I learned to play Ping-Pong, I was able to transfer some of those skills to help me in learning tennis and badminton. Knowing how to play checkers helped me in learning how to play chess. When I was a toddler, it took me a while to learn how to twist the doorknob in my room, but once I had mastered that skill, my abilities quickly generalized to most any kind of doorknob.

Humans exhibit this kind of transfer from one task to another seemingly effortlessly; our ability to generalize what we learn is a core part of what it means for us to *think*. Thus, in human-speak, we might say that another term for transfer learning is, well, *learning*.

In stark contrast with humans, most "learning" in current-day AI is not transferable between related tasks. In this regard, the field is still far from what Hassabis calls "general AI." While the topic of transfer learning is one of the most active areas of research for machine-learning practitioners, progress on this front is still nascent.[3]

"Without Human Examples or Guidance"

Unlike supervised learning, reinforcement learning holds the promise of programs that can truly learn on their own, simply by performing actions in their "environment" and observing the outcome. DeepMind's most important claim about its results, especially on AlphaGo, is that the work has delivered on that promise: "Our results comprehensively demonstrate that a pure reinforcement learning approach is fully feasible, even in the most challenging of domains: it is possible to train to superhuman level, without human examples or guidance, given no knowledge of the domain beyond basic rules."[4]

We have the claim. Now let's look at the caveats. AlphaGo (or more precisely, the AlphaGo Zero version) indeed didn't use any human examples in its learning, but human "guidance" is another story. A few aspects of human guidance that were critical to its success include the specific architecture of its convolutional neural network, the use of Monte Carlo tree search, and the setting of the many hyperparameters that both of these entail. As the psychologist and AI researcher Gary Marcus has pointed out, none of these crucial aspects of AlphaGo were "learned from the data, by pure reinforcement learning. Rather, [they were] built in innately . . . by DeepMind's programmers."[5] DeepMind's Atari game-playing programs were actually better examples of "learning without human guidance" than AlphaGo, because unlike the latter they were not provided with the rules of their game (for example, that the goal in *Breakout* is to destroy bricks) or even a concept of the "objects" relevant to the game (for example, *paddle* or *ball*) but learned exclusively from the screen pixels.

The Most Challenging of Domains

One additional aspect of DeepMind's statement needs to be explored: "even in the most challenging of domains." How can we assess how challenging a domain is for AI? As we've seen, many things we humans consider quite easy (for example, describing the contents of a photo) are extremely challenging for computers. Conversely, many things we humans would find terrifically challenging (for example, correctly multiplying two fifty-digit numbers), computers can do in a split second with a one-line program.

One way to assess the challenge of a domain for computers is to see how well very simple algorithms perform on it. In 2018, a group of researchers at Uber AI Labs found that some relatively simple algorithms nearly matched (and sometimes outperformed) DeepMind's deep Q-learning method on several Atari video games. The most surprising good performer was "random search": instead of training a Deep Q-Network by reinforcement learning over many episodes, one can simply try out many different convolutional neural networks with *randomly* chosen weights.[6] That is, there is no learning whatsoever, except via random trial and error.

You'd think that a network with random weights would perform abominably on an Atari video game. Indeed, most such networks are terrible players. But the Uber researchers kept trying new random-weight networks, and eventually (in less time than it took to train a Deep Q-Network) they found networks that performed nearly as well as or even better than networks trained by deep Q-learning on five out of the thirteen games they tested. Another relatively simple algorithm, a so-called genetic algorithm,[7] outperformed deep Q-learning on seven out of thirteen games. It's hard to know what to say about these results, except that it's possible that the Atari game domain is not as challenging for AI as people originally thought.

I haven't heard of anyone trying a similar random search for network weights for Go. I'd be very surprised if that worked at all. Given the long history of attempts to build computer Go players, I'm convinced that Go counts as a genuinely challenging domain for AI. However, as Gary Marcus pointed out, there are many games humans play that are even more challenging for AI than Go. One striking example Marcus gives is charades,[8] which, if you think about it, requires sophisticated visual, linguistic, and social understanding far beyond the abilities of any current AI system. If you could build a robot that could play charades as well as, say, a six-year-old child, then I think you could safely say that you had conquered several of the "most challenging of domains" for AI.

What Did These Systems Learn?

Like other applications of deep learning, it's hard to interpret what the neural networks used in these game-playing systems have actually learned. In reading the sections above, you might have noticed some subtle anthropomorphism creeping into my descriptions: for example, I said, "DeepMind's *Breakout* player discovered the strategy of tunneling through the bricks."

It's dangerously easy, for me as much as anyone, to slip into this kind of language when talking about the behavior of AI systems. However, our language often carries unconscious assumptions that may not hold for these programs. Did DeepMind's *Breakout* player actually discover the concept of *tunneling*? Gary Marcus reminds us that we need to be careful here:

> The system has learned no such thing; it doesn't really understand what a tunnel, or what a wall is; it has just learned specific contingencies for particular scenarios. Transfer tests—in which the deep reinforcement learning system is confronted with scenarios that differ in minor ways from the ones on which the system was trained—show that deep reinforcement learning's solutions are often extremely superficial.[9]

Marcus is referring to a few studies that tried to probe how well deep Q-learning systems can transfer what they learned, even to very small variations of the same game. For example, one group of researchers studied a system similar to DeepMind's *Breakout* player. They found that once the player is trained to "superhuman" level, if the paddle's position on the screen is shifted up by a few pixels, the system's performance plummets.[10] This hints that the system has not even learned the basic concept of *paddle*. Another group showed that for a deep Q-learning system trained on the game *Pong*, if the screen's background color is changed, the system's performance decreases significantly.[11] Moreover, in each case the system needs many episodes of retraining to adapt to the variation.

These are just two examples of deep Q-learning's inability to generalize, which contrasts strikingly with human intelligence. I don't know of any study that probed the concept of tunneling in DeepMind's *Breakout* player, but I'd guess that the system couldn't generalize to, say, tunneling *down* or *sideways*, without considerable retraining. As Marcus notes, while *we humans* attribute to the program a certain understanding of what we consider basic concepts (for example, *wall, ceiling, paddle, ball, tunneling*), the program actually has no such concepts:

> These demonstrations make clear that it is misleading to credit deep reinforcement learning with inducing concepts like *wall* or *paddle*; rather, such remarks are what comparative (animal) psychology sometimes call overattributions. It's not that the Atari system genuinely learned a concept of wall that was robust but rather the system superficially approximated breaking through walls within a narrow set of highly trained circumstances.[12]

Similarly, while AlphaGo exhibited miraculous "intuition" in playing Go, the system doesn't have any mechanisms, as far as I can tell, that would allow it to generalize its Go-playing abilities, even to, say, a smaller or differently shaped Go board, without restructuring and retraining its Deep Q-Network.

In short, while these deep Q-learning systems have achieved superhuman performance in some narrow domains, and even exhibit what resembles "intuition" in these domains, they are lacking something absolutely fundamental to human intelligence. Whether it is called abstraction, domain generalization, or transfer learning, imbuing systems with this ability is still one of AI's most important open problems.

There's another reason to suspect that these systems are not learning humanlike concepts or *understanding* their domains in the way humans do: like supervised-learning systems, these deep Q-learning systems are vulnerable to adversarial examples of the kind I described in chapter 6. For example, one research group showed that it's possible to make specific minuscule changes to the pixels in an Atari game-playing program's input—changes that are imperceptible to humans but that significantly damage the program's ability to play the game.

How Intelligent Is AlphaGo?

Here's something we must keep in mind when thinking about games like chess and Go and their relation to human intelligence. Consider the reasons many parents encourage their kids to join the school chess club (or in some places the Go club) and would much rather see their kids playing chess (or Go) than sitting at home watching TV or playing video games (sorry, Atari). It's because people believe that games like chess or Go teach children (or anyone) how to *think* better: how to think logically, reason abstractly, and plan strategically. These are all capabilities that will carry over into the rest of one's life, general abilities that a person will be able to use in all endeavors.

But AlphaGo, in spite of the millions of games it has played during its training, has not learned to "think" better about anything except the game of Go. In fact, it has no ability to think about anything, to reason about anything, to make plans about anything, except Go. As far as I know, none of the abilities it has learned are general in any way; none can be transferred to any other task. AlphaGo is the ultimate idiot savant.

It's certainly true that the deep Q-learning method used in AlphaGo can be used to learn other tasks, but the system itself would have to be wholly retrained; it would have to start essentially from scratch in learning a new skill.

This brings us back to the "easy things are hard" paradox of AI. AlphaGo was a great achievement for AI; learning largely via self-play, it was able to definitively defeat one of the world's best human players in a game that is considered a paragon of intellectual prowess. But AlphaGo does not exhibit human-level intelligence as we generally define it, or even arguably any real intelligence. For humans, a crucial part of intelligence is, rather than being able to learn any particular skill, being able to *learn to think* and to then apply our thinking flexibly to whatever situations or challenges we encounter. This is the true skill we want our children to learn when they play chess or Go. It may sound strange to say, but in this way the lowliest kindergartner in the school chess club is smarter than AlphaGo.

From Games to the Real World

Finally, let's consider Demis Hassabis's statement that the ultimate goal of these demonstrations on games is to "use them so they apply to real-world problems and have a huge impact on things like healthcare and science." I think it's very possible that DeepMind's work on reinforcement learning may eventually have the kinds of impacts Hassabis is aiming for. But there's a long way to go from games to the real world.

The need for transfer learning is one obstacle. But there are additional reasons that it will be difficult to extend reinforcement learning's success in games to the real world. Games such as *Breakout* and Go are ideally suited for reinforcement learning because they have clear rules, straightforward reward functions (for example, rewards for points gained or for winning), and relatively few possible actions (moves). Moreover, the players have access to "perfect information": all the components of the game are visible at all times to the players; there are no hidden or uncertain parts of a player's "state."

The real world doesn't come so cleanly delineated. Douglas Hofstadter has pointed out that the very notion of a clearly defined "state" isn't at all realistic. "If you look at situations in the world, they don't come framed, like a chess game or a Go game. . . . A situation in the world is something that has no boundaries at all; you don't know what's in the situation, what's out of the situation."[13]

As an example, consider using reinforcement learning to train a robot to perform a very useful real-world task: take the dirty dishes stacked in the sink and put them in the dishwasher. (Oh, the harmony such a robot would bring to my family!) How should we define the robot's "state"? Would it involve everything in its visual field? The contents of the sink? The contents of the dishwasher? How about the dog, coming over to lick the dishes, who needs to be shooed away? However we define its state, the robot would need to be able to identify different objects—for example, recognizing a plate (which should go on the bottom rack of the dishwasher), a coffee mug (which should go on the top rack), or a sponge (which doesn't go in the dishwasher at all). As we've seen, object recognition by computers is as yet far from perfect. In addition, the robot would have to reason about objects that it can't see—perhaps pots and pans hidden at the bottom of the sink. The robot would also need to learn to pick up different objects and place them (carefully!) in appropriate slots. All this would require learning to choose among a multitude of possible actions involving the robot's body placement, its grasping "fingers," its motors controlling the movement of objects from the sink to the correct dishwasher slot, and so on.[14]

DeepMind's game-playing agents required millions of iterations of training. If we don't want millions of broken dishes, we'd have to train our robot in simulation. Games are very fast and accurate to simulate on a computer; there's no actual moving of pieces or actual balls bouncing off paddles or actual bricks exploding. But simulating a dishwasher-loading robot is not so easy. The more realistic the simulation, the slower it is to run on a computer, and even with a very fast computer it's enormously difficult to incorporate all the physical forces and other aspects of dish loading accurately into the simulation. And then there's that pesky dog, as well as all the other unpredictable aspects of the real world; how do we figure out what needs to be in the simulation and what can safely be ignored?

All these issues led Andrej Karpathy, Tesla's director of AI, to note that, for real-world tasks like this, "basically every single assumption that Go satisfies and that AlphaGo takes advantage of are violated, and any successful approach would look extremely different."[15]

No one knows what that successful approach would be. Indeed, the field of deep reinforcement learning is still quite young. The results I described in this chapter can be seen as a proof of principle: the combination of deep networks and Q-learning works surprisingly well in some very interesting, albeit narrow, domains, and while my discussion has highlighted some of the current limitations of the field, many people are working on extending reinforcement learning to apply more generally. DeepMind's game-playing programs in particular have ignited tremendous new interest and enthusiasm in the field; in fact, deep reinforcement learning was named one of 2017's "10 Breakthrough Technologies" by MIT's *Technology Review* magazine. In years to come, as reinforcement learning matures, I'll be eagerly awaiting a dishwasher-loading robot that learns on its own, and maybe plays both soccer and Go in its spare time.

Part IV

·

Artificial

·

Intelligence Meets

·

Natural Language

·

11

Words, and the

Company They Keep

It's time for a story.

The Restaurant

A man went into a restaurant and ordered a hamburger, cooked rare.
When it arrived, it was burned to a crisp. The waitress stopped by the
man's table. "Is the burger okay?" she asked. "Oh, it's just great," the man
said, pushing back his chair and storming out of the restaurant without
paying. The waitress yelled after him, "Hey, what about the bill?" She
shrugged her shoulders, muttering under her breath, "Why is he so bent
out of shape?"[1]

Now let me ask you a question: Did the man eat the hamburger?

I'm guessing that you're quite confident of your answer, even though the story doesn't directly address this question. It's easy, at least for us as humans, to read between the lines. After all, understanding language—including the parts that are left *unsaid*—is a fundamental part of human intelligence. It's no accident that Alan Turing framed his famous "imitation game" as a contest involving the generation and understanding of language.

This part of the book deals with *natural-language processing*, which means "getting computers to deal with human language." (In AI-speak, "natural" means "human.") Natural-language processing (abbreviated NLP) includes topics such as speech recognition, web search, automated question answering, and machine translation. Similar to what we've seen in previous chapters, deep learning has been the driving force behind most of the recent advances in NLP. I'll describe some of these advances, using the "Restaurant" story to illustrate a few of the major challenges machines face when it comes to using and understanding human language.

The Subtlety of Language

Suppose we want to create a program that can read a passage and answer questions about it. Question-answering systems are a central focus of current NLP research, because people want to use natural language to interact with computers (think about Siri, Alexa, Google Now, and other "virtual assistants"). However, in order to answer questions about a text such as the "Restaurant" story, a program would require sophisticated linguistic skills as well as substantial knowledge about the way the world works.

Did the man eat the hamburger? To answer this confidently, a hypothetical program would need to know that hamburgers belong to the category "food," and that foods can be *eaten*. The program should know that going into a restaurant and ordering a hamburger usually means that you plan to eat the hamburger. Moreover, in a restaurant, once your order has arrived, it is available to be eaten. A program would need to know that when a person orders a hamburger "cooked rare," the person generally doesn't want to eat it if it has been "burned to a crisp." The program should recognize that when the man says, "Oh, it's just great," he is being sarcastic, and that "it" refers to the "burger," which is another word for "hamburger." The program would need to surmise that if you "storm" out of a restaurant without paying, it's likely that you haven't eaten your meal.

It's mind-boggling to think of all the background knowledge the program would need in order to give confident answers to basic questions about the story. Did the man leave the waitress a tip? The program would need to know about the custom of tipping in restaurants and its purpose of rewarding good service. Why did the waitress say, "What about the bill"? The program needs to figure out that by "bill," the waitress is referring not to, say, the beak of a bird, or a banknote, or a written piece of legislation, but to the charge for the man's meal. Did the waitress know that the man was angry? The program has to determine that in asking "Why is he so bent out of shape?" "he" refers to the man, and "bent out of shape" is an idiom that means "upset and angry." Did the waitress know *why* the man left the restaurant? It would help if our program knew that the gesture "shrugging her shoulders" suggests that the waitress didn't understand why he stormed out.

Thinking about what our hypothetical program would need to know reminds me of trying to answer the endless questions my children would ask when they were very young. Once, when my son was four years old, I took him with me to go to the bank. He asked a simple question: "What's a bank?" My answer prompted a seemingly endless cascade of "why" questions. "Why do people use money?" "Why do people want to have a lot of money?" "Why can't people keep all their money at home?" "Why can't I make my own money?" All good questions, but hard to answer without having to explain all sorts of things that lie beyond a four-year-old's experience.

The situation is much more extreme for machines. A child who hears the "Restaurant" story already has well-grounded concepts such as *person*, *table*, and *hamburger*. Children have basic common sense, knowing, for example, that when the man walks out of the restaurant, he is no longer *inside* the restaurant, but the tables and chairs are probably still there. Or when the hamburger "arrived," someone brought it to the man's table (it didn't arrive on its own). Today's machines lack the detailed, interrelated concepts and commonsense knowledge that even a four-year-old child brings to understanding language.

It should come as no surprise, then, that using and understanding natural language are among AI's most difficult challenges. Language is inherently ambiguous, is deeply dependent on context, and assumes a great deal of background knowledge common to the communicating parties. As with other areas of AI, the first several decades of NLP research focused on symbolic *rule-based* approaches—that is, programs that were given grammatical and other linguistic rules and applied these rules to input sentences. These approaches did not work very well; it seems to be impossible to capture the subtleties of language by applying a set of explicit rules. In the 1990s, rule-based NLP approaches were overshadowed by more successful *statistical* approaches, in which massive data sets were employed to train machine-learning algorithms. Most recently, this statistical data-driven approach has focused on deep learning. Can deep learning, along with big data, produce machines that can flexibly and reliably deal with human language?

Speech Recognition and the Last 10 Percent

Automated speech recognition—the task of transcribing spoken language into text in real time—was deep learning's first major success in NLP, and I'd venture to say that it is AI's most significant success to date in any domain. In 2012, at the same time that deep learning was revolutionizing computer vision, a landmark paper on speech recognition was published by research groups at the University of Toronto, Microsoft, Google, and IBM.[2] These groups had been developing deep neural networks for various aspects of speech recognition: recognizing phonemes from acoustic signals, predicting words from combinations of phonemes, predicting phrases from combinations of words, and so on. According to a Google speech-recognition expert, the use of deep networks resulted in the "biggest single improvement in 20 years of speech research."[3] The same year, a new deep-network speech-recognition system was released to customers on Android phones; two years later it was released on Apple's iPhone, with one Apple engineer commenting, "This was one of those things where the jump [in performance] was so significant that you do the test again to make sure that somebody didn't drop a decimal place."[4]

If you yourself happened to use any kind of speech-recognition technology both before and after 2012, you will have also noticed a very sharp improvement. Speech recognition, which before 2012 ranged from horribly frustrating to moderately useful, suddenly became very nearly perfect in some circumstances. I am now able to dictate all of my texts and emails on my phone's speech-recognition app; just a few moments ago, I read the "Restaurant" story to my phone, using my normal speaking speed, and it correctly transcribed every word.

What's stunning to me is that speech-recognition systems are accomplishing all this without any understanding of the *meaning* of the speech they are transcribing. While the speech-recognition system on my phone can transcribe every word of my "Restaurant" story, I guarantee you that it doesn't understand a thing about it, or about anything else. Many people in AI, myself included, had previously believed that AI speech recognition would never reach such a high level of performance without actually understanding language. But we've been proven wrong.

This being said, automated speech recognition is still not at "human level," contrary to some reports in the media. Background noise can significantly hurt the accuracy of these systems; they're much less effective inside a moving car than in a quiet room. In addition, these systems are occasionally thrown off by unusual words or phrases in a way that highlights their lack of understanding of the speech they are transcribing. For example, I said, "Mousse is my favorite dessert," but my (Android) phone transcribed it as "Moose is my favorite dessert." I said, "The bareheaded man needed a hat," but my phone transcribed it as "The bear headed man needed a hat." It's not hard to find sentences that will confuse a speech-recognition system. However, for everyday speech in a quiet environment, I'd guess that the accuracy of such systems—measured by correct words—is probably around 90 to 95 percent of humans' accuracy.[5] If you add noise or other complications, the accuracy goes down considerably.

There's a famous rule of thumb in any complex engineering project: the first 90 percent of the project takes 10 percent of the time and the last 10 percent takes 90 percent of the time. I think that some version of this rule applies in many AI domains (hello, self-driving cars!) and will end up being true in speech recognition as well. The last 10 percent includes dealing not only with noise, unfamiliar accents, and unknown words but also with the fact that the ambiguity and context sensitivity of language can impinge on interpreting speech. What's needed to power through that last stubborn 10 percent? More data? More network layers? Or, dare I ask, will that last 10 percent require an actual *understanding* of what the speaker is saying? I'm leaning toward this last one, but I've been wrong before.

Speech-recognition systems are quite complicated; several different kinds of processing are needed to go from sound waves to sentences. Current state-of-the-art speech-recognition systems integrate several different components, including multiple deep neural networks.[6] Other NLP tasks, such as language translation or question answering, seem simpler at first glance: the input and output both consist of words. However, deep learning's data-driven approach hasn't produced the same kind of progress in these areas as it did in speech recognition. Why not? To answer, let's look at a few examples of how deep learning has been applied to important NLP tasks.

Classifying Sentiment

As a first example, let's look at the area called sentiment classification. Consider these short reviews of the movie *Indiana Jones and the Temple of Doom*:[7]

"The plot is heavy and the sense of humor is largely missing."

"A little too dark for my taste."

"It felt as if the producers tried to make it as disturbing and horrific as they possibly could."

"Temple of Doom's character development and humor is intensely subpar."

"The tone is kind of weird and it has a lot of humor that wasn't working for me."

"Without any of the charm or wit that is embodied in the others in this series."

In each case, did the reviewer like the movie?

There is big money in using machines to answer such a question. An AI system that could accurately classify a sentence (or longer passage) as to its *sentiment*—positive, negative, or some other degree of opinion—would be solid gold to companies that want to analyze customers' comments about their products, find new potential customers, automate product recommendations ("people who liked X also like Y"), or selectively target their online advertisements. Data on what movies, books, or other merchandise a person likes or dislikes can be surprisingly (and perhaps scarily) useful in predicting that person's future purchases. What's more, such information may have predictive power about other aspects of a person's life, such as likely voting patterns and responsiveness to certain types of news stories or political ads.[8] Furthermore, there have been several efforts, with varying success, to apply "sentiment mining" of, say, economics-related tweets on Twitter to predict stock prices and election outcomes.

Putting aside the ethics of these applications of sentiment analysis, let's focus on *how* AI systems might be able to classify the sentiment of sentences like the ones above. While it's quite easy for humans to see that these mini-reviews are all negative, getting a program to do this kind of classification in a general way is much harder than it might seem at first glance.

Some early NLP systems looked for the presence of individual words or short sequences of words as indications of the sentiment of a text. For example, you might expect words such as *dark, weird, heavy, disturbing, horrific, lacking,* and *missing,* or sequences such as *wasn't working, without any, a little too,* as indicating negative sentiment in movie reviews. In some cases this works, but in many cases such sequences can be found in positive reviews as well. Here are a few examples:

> "Despite the heavy subject matter, there's enough humor to keep it from becoming too dark."

> "There's nothing here that is disturbing or horrific as some people have suggested."

> "I was a little too young to see this terrific movie when it first came out."

> "If you don't see it, you'll be missing out!"

Looking at single words or short sequences in isolation is generally not sufficient to glean the overall sentiment; it's necessary to capture the semantics of words in the context of the whole sentence.

Soon after deep networks started to excel in computer vision and speech recognition, NLP practitioners experimented with applying them to sentiment analysis. As usual, the idea is to train the network on many human-labeled examples of sentences with both positive and negative sentiment and have the network itself learn useful *features* that allow it to output a classification confidence for "positive" or "negative" on a new sentence. But first, how can we get a neural network to process a sentence?

Recurrent Neural Networks

Processing a sentence or passage requires a different type of neural network from those I have described in previous chapters. Recall, for example, the convolutional neural network from chapter 4 that classified an image as "dog" or "cat." There, the network's inputs were the pixel intensities of a fixed-size image (larger or smaller images had to be scaled to the proper size). In contrast, sentences consist of *sequences* of words and do not have a fixed length. Thus, we need a way for a neural network to process variable-length sentences.

Applying neural networks to tasks involving ordered sequences such as sentences goes back to the 1980s, with the introduction of *recurrent neural networks* (RNNs), which were inspired, of course, by ideas on how the brain interprets sequences. Imagine that you are asked to read the review "A little too dark for my taste" and classify it as having positive or negative sentiment. You read the sentence left to right, one word at a time. As you read it, you start to form impressions of its sentiment, which become further supported as you finish reading the sentence. At this point, your brain has some kind of representation of the sentence in the form of neural activations, which allow you to confidently state whether the review is positive or negative.

Recurrent neural networks are loosely inspired by this sequential process of reading a sentence and creating a representation of it in the form of neural activations. Figure 32 compares the structures of a traditional neural network and a recurrent neural network. For simplicity, each network has two units (white circles) in the hidden layer and one unit in the output layer. In both networks, the input has connections to the hidden units,

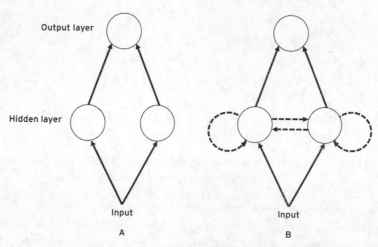

Output layer

Hidden layer

Input
A

Input
B

FIGURE 32: *A*, illustration of a traditional neural network; *B*, illustration of a recurrent neural network, in which the activations of the hidden units at a given time step are fed back at the next time step

and each hidden unit has a connection to the output unit (solid arrows). The key difference for the RNN is that its hidden units have additional "recurrent" connections; each hidden unit has a connection to itself and to the other hidden unit (dashed arrows). How does this work? Unlike a traditional neural network, an RNN operates over a series of time steps. At each time step, the RNN is fed an input and computes the activation of its hidden and output units just as does a traditional neural network. But in an RNN each hidden unit computes its activation based on both the input and the activations of the hidden units *from the previous time step*. (At the first time step, these recurrent values are set to 0.) This gives the network a way to interpret the words it "reads" while remembering the context of what it has already "read."

The best way to understand how RNNs work is to visualize the network's operation over time, as in figure 33, which shows the RNN from figure 32 over eight time steps. To simplify the illustration, I represent all the recurrent connections in the hidden layer as a single dashed arrow from one time step to the next. At each time step, the hidden units' activations constitute the network's encoding of the partial sentence it has

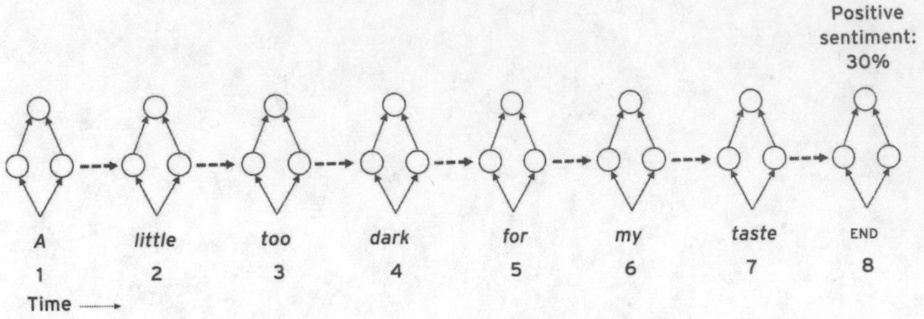

FIGURE 33: The recurrent neural network of figure 32, operating over eight time steps

seen so far. The network keeps refining that encoding as it continues to process words. After the last word in the sentence, the network is given a special END symbol (similar to a period), which tells the network that the sentence is finished. Note that the END symbol is appended by humans to each sentence before the text is fed to the network.

At each time step, the output unit in this network processes the hidden units' activations (the "encoding") to give the network's confidence that the input sentence (that is, the part of the sentence given to the network up to that time step) has a *positive* sentiment. When applying the network to a given sentence, we can ignore this output until the end of the sentence has been reached. At this point, the hidden units encode the entire sentence, and the output unit gives the network's final confidence (here, 30 percent for positive sentiment or, equivalently, 70 percent for negative sentiment).

Because the network stops encoding the sentence only when it encounters the END symbol, the system can in principle encode sentences of any length into a fixed-length set of numbers—the activations of the hidden units. For obvious reasons, this kind of neural network is often called an encoder network.

Given a set of sentences that humans have labeled as "positive" or "negative" in sentiment, the encoder network can be trained from these examples via back-propagation. But there's one thing I haven't explained yet. Neural networks require their inputs to be *numbers*.[9] What is the best way to encode the input words as numbers? Answering this question has led to one of the most important advances in natural-language processing in the last decade.

A Simple Scheme for Encoding Words as Numbers

Before explaining possible schemes for encoding words as numbers, I need to define the notion of a neural network's *vocabulary*. The vocabulary is the set of all words that the network will be able to accept as inputs. Linguists estimate that on the order of ten thousand to thirty thousand words are needed for a reader to deal with most English texts, depending on how you count; for example, you might group together *argue*, *argues*, *argued*, and *arguing* as one "word." The vocabulary can also include common two-word phrases, for example, *San Francisco* or *Golden Gate*, by counting them as a single word.

As a concrete example, let's assume that our network will have a twenty-thousand-word vocabulary. The simplest possible scheme for encoding words as numbers is to assign each word in the vocabulary an arbitrary number between 1 and 20,000. Then give the neural network 20,000 inputs, one per word in the vocabulary. At each time step, only one of those inputs—the one corresponding to the actual input word—will be "switched on." For example, say that the word *dark* has been given the number 317. Then, if we want to input *dark* to the network, we set input 317 to have value 1, and all the other 19,999 inputs to have value 0. In the NLP field, this is called a one-hot encoding: at each time step, only one of the inputs—the one corresponding to the word being fed to the network—is "hot" (non-0).

The one-hot encoding used to be a standard way to input words to neural networks. But it has a problem: an arbitrary assignment of numbers to words doesn't capture any relationships among words. Suppose that the network has learned from its training data that the phrase "I hated this movie" has negative sentiment. Now suppose that the network is given the phrase "I abhorred this flick," but it has not encountered *abhorred* or *flick* in its training data. The network wouldn't have any way to determine that the meanings of the two phrases are the same. Suppose further that the network has learned that the phrase "I laughed out loud" is associated with positive reviews, and then it encounters the novel phrase "I appreciated the humor." The network wouldn't be able to recognize the close (though not exactly identical) meanings of these two phrases. The inability to capture semantic relationships among words and phrases is a major reason why neural networks using one-hot encodings often don't work very well.

The Semantic Space of Words

The NLP research community has proposed several methods for encoding words in a way that would capture such semantic relationships. All of these methods are based on the same idea, which was expressed beautifully by the linguist John Firth in 1957: "You shall know a word by the company it keeps."[10] That is, the meaning of a word can be defined in terms of other words it tends to occur with, and the words that tend to occur with those words, and so on. *Abhorred* tends to occur in the same contexts as *hated*. *Laughed* tends to occur with the same words that *humor* finds in its company.

In linguistics, this idea is known more formally as distributional semantics. The underlying hypothesis of distributional semantics is that "the degree of semantic similarity between two linguistic expressions A and B is a function of the similarity of the linguistic contexts in which A and B can appear."[11] Linguists often make this more concrete via the idea of a "semantic space." Figure 34A illustrates a two-dimensional semantic space of words in which words with similar meanings are located closer to one another. But one quickly sees that because words can have many dimensions of meaning, their semantic space must have more dimensions as well. For example, the word *charm* is close to *wit* and *humor*, but in a different context *charm* is close to *bracelet* and *jewelry*. Similarly, the word *bright* is close to both the *light* cluster and the *happy* cluster but also has an alternative (though related) meaning that is close to *smart*, *intelligent*, and *clever*. It would be helpful to have a third dimension, coming toward you out of the page, to place these words at the proper distance from one another. Along one dimension, *charm* is near *wit*; along another, it is near *bracelet*. But *charm* should also be close to *lucky* (whereas *bracelet* is not). We need more dimensions! We humans have trouble picturing a space of more than three dimensions, but the semantic space of words might actually require many dozens if not hundreds of dimensions.

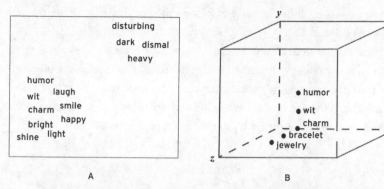

FIGURE 34: *A*, illustration of two word-clusters in a semantic space in which words with similar meanings are located close to one another; *B*, a three-dimensional semantic space in which words are plotted as points

When we're talking about semantic *spaces* with multiple *dimensions*, we find ourselves in the realm of geometry. Indeed, NLP practitioners often frame the "meaning" of words in terms of geometric concepts. For example, figure 34B shows a three-dimensional space, with x-, y-, and z-axes, along which words can be placed. Each word is identified with a point (black circle), defined by three coordinates—that is, the x, y, and z locations of the point. The semantic distance between two words is equated with the geometric distance between points on this plot. You can see that *charm* is now close to both *wit* and *humor* and to *bracelet* and *jewelry*, but along different dimensions. In NLP, people use the term *word vector* to refer to the coordinates of a particular word in such a semantic space. In mathematics, *vector* is just a fancy term for the coordinates of a point.[12] For example, suppose that *bracelet* happens to be located at coordinates (2, 0, 3); this list of three numbers is its word vector in this three-dimensional space. Note that the number of dimensions in a vector is simply the number of coordinates.

The idea here is that once all the words in the vocabulary are properly placed in the semantic space, the *meaning* of a word can be represented by its location in this space—that is, by the coordinates defining its word vector. And what is a word vector good for? It turns out that using word vectors as numerical inputs to represent words, as opposed to the simple one-hot scheme I sketched above, greatly improves the performance of neural networks in NLP tasks.

How do we actually obtain all the word vectors corresponding to words in a vocabulary? Is there an algorithm that will properly place all the words in our network's vocabulary in a semantic space in order to best capture the many dimensions of each word's meaning? A lot of important work in NLP has gone into solving this exact problem.

Word2Vec

Many solutions have been suggested for the problem of placing words in a geometric space, some going back to the 1980s, but today's most widely adopted method was proposed in 2013 by researchers at Google.[13] The researchers called their method "word2vec" (shorthand for "word to vector"). The word2vec method uses a traditional neural network to automatically learn word vectors for all the words in a vocabulary. The Google researchers used part of the company's vast store of documents to train their network; once training was completed, the Google group saved and published all the resulting word vectors on a web page for anyone to download and use as input to natural-language processing systems.[14]

The word2vec method embodies the notion of "you shall know a word by the company it keeps." To create the training data for the word2vec program, the Google group started by taking a massive set of documents from the Google News service. (In modern NLP, nothing beats having "big data" lying around!) The training data for the word2vec program consisted of a collection of pairs of words, where each word in the pair had occurred near the other word in the pair somewhere in the Google News documents. To make the process work better, extremely frequent words such as *the*, *of*, and *and* were discarded.

As a concrete example, assume that the words of each pair occur immediately adjacent to each other in a sentence. In this case, the sentence "a man went into a restaurant and ordered a hamburger" first would be transformed into "man went into restaurant ordered hamburger." This would yield the following pairs: (man, went), (went, into), (into, restaurant), (restaurant, ordered), (ordered, hamburger), plus the reverse of all

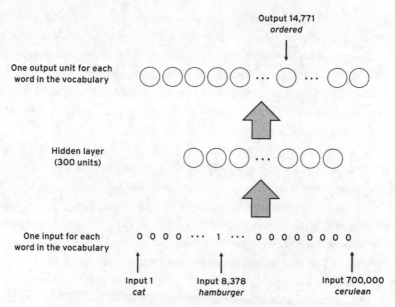

FIGURE 35: Illustration of the word2vec neural network, given the word pair (hamburger, ordered)

the pairs—for example, (hamburger, ordered). The idea is to train the word2vec network to predict what words are likely to be paired with a given input word.

Figure 35 illustrates the word2vec neural network.[15] This network actually uses the one-hot encoding I described above. In figure 35, there are 700,000 input units; this is close to the vocabulary size used by the Google researchers. Each input corresponds to a word in the vocabulary. For example, the first input here corresponds to the word *cat*, the 8,378th input corresponds to *hamburger*, and the 700,000th input corresponds to *cerulean*. I just made up these numbers; the actual ordering doesn't matter. Similarly, there are 700,000 output units, each corresponding to a word in the vocabulary, and a relatively small hidden layer of 300 units. The large gray arrows indicate that each input has a weighted connection to each hidden unit, and each hidden unit has a weighted connection to each output unit.

The Google researchers trained their network on billions of word pairs collected from Google News articles. Given a word pair such as (hamburger, ordered), the input corresponding to the first word in the pair (hamburger) is set to 1; all other inputs are set to 0. During training, each output unit's activation is interpreted to be the network's confidence that the corresponding word in the vocabulary has occurred adjacent to the input word. Here, the *correct* output activations would assign high confidence to the second word in the pair (ordered).

After the training is completed, one can extract the learned *word vector* for any word in the vocabulary. Figure 36 illustrates how this is done. The figure shows the weighted connections between one input (corresponding to the word *hamburger*) and the three hundred hidden units. These weights, which have been learned from the training data, have captured information about the contexts in which the corresponding word is used. These three hundred weight values are the components of the word vector assigned to the given word. (The connections from the hidden units to the outputs are completely ignored in this process; all the necessary information resides in the input-to-hidden-layer weights.) Thus, the word vectors learned by this network have three hundred dimensions. The collection of word vectors for all words in the vocabulary constitutes the learned "semantic space."

Here's how you can visualize this three-hundred-dimensional semantic space in your head. Just think of the three-dimensional plot from figure 34 and then try to visualize a similar plot with a hundred times as many dimensions, and with seven hundred thousand words plotted, each with three hundred coordinates. Just kidding! It's impossible to visualize such a thing.

What do these three hundred dimensions represent? If we ourselves were three-hundred-dimensional creatures who had the brains to visualize such a space, we'd see that any given word is close to other related words across many meanings. For example, the vector for *hamburger* is close to the vector for *ordered*; it is also close to the vectors for *burger, hot dog, cow, eat*, and so on. *Hamburger* is also close to *dinner* even if it has never been seen in a pair with *dinner*; this is because *hamburger* is close to words that are also close to *dinner* in similar contexts. If the network sees word pairs from "I ate a hamburger for lunch" as well as from "I devoured a hot

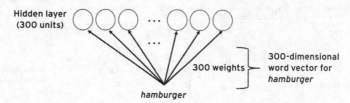

dog for dinner," and if *lunch* and *dinner* also appear close together in some training sentences, then the system can learn that *hamburger* and *dinner* should also be close.

Let's remember that the goal of this whole process is to find a numerical representation—a vector—for each word in the vocabulary, one that captures something of the semantics of the word. The hypothesis is that using such word vectors will result in high-performing neural networks for natural-language processing tasks. But to what extent does the "semantic space" created by word2vec actually capture word semantics?

This question is hard to answer, because we can't visualize the three-hundred-dimensional semantic space learned by word2vec. However, we can do a few things to glimpse into this space. The simplest approach is to take a given word and find the words that have ended up closest to it in the semantic space, by looking at the distances between word vectors. For example, after the network has been trained, the closest words to *France* include *Spain, Belgium, Netherlands, Italy, Switzerland, Luxembourg, Portugal, Russia, Germany,* and *Catalonia*.[16] The word2vec algorithm wasn't told the concept of *country* or *European country*; these are just the words that appear in the training data in similar contexts to *France*, the way *hamburger* and *hot dog* do in my example above. Indeed, if I ask for the closest words to *hamburger*, the list includes *burger, cheeseburger, sandwich, hot dog, taco,* and *fries*.[17]

We can also look at more complex relationships that result from the network's training. The Google researchers who created word2vec observed that in the word vectors created by their network, the distance between the

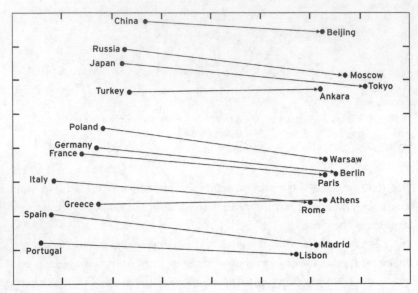

FIGURE 37: Two-dimensional representation of distances between word vectors for countries and word vectors for their capital cities

word for a country and the word for that country's capital is approximately the same for many countries. This is illustrated in figure 37, which shows a two-dimensional representation of these distances. Again, the system wasn't given the notion of a "capital" of a country; these relationships simply emerged from the network's training on billions of word pairs.

This kind of regularity gave people the idea that word2vec could "solve" analogy problems such as *"Man* is to *woman* as *king* is to _____." Just take the word vector for *woman*, subtract the word vector for *man*, and add the result to the word vector for *king*.[18] Then find the word vector in the space that is closest to the result. Yup, it's *queen*. In my experimenting with an online word2vec demonstration,[19] this method often produces some very good results (*"Dinner* is to *evening* as *breakfast* is to *morning"*), but just as often it is cryptic (*"Thirsty* is to *drink* as *tired* is to *drunk"*) or nonsensical (*"Fish* is to *water* as *bird* is to *hydrant"*).

Such properties of learned word vectors are intriguing and show that some relationships are captured. But will these properties make word vectors generally useful in NLP tasks? The answer seems to be a resounding "yes." Nowadays, virtually all NLP systems use word vectors of one sort or another (word2vec is only one flavor) as the way to input words.

Here's an analogy for you: to a person with a hammer, everything looks like a nail; to an AI researcher with a neural network, everything looks like a vector. It occurred to many people that the word2vec trick could be played not only with words but with whole sentences as well. Why not encode an entire sentence as a vector in the same way that words are encoded, using sentence pairs instead of word pairs in training? Wouldn't something like this capture semantics better than simply a set of word vectors? Indeed, several groups have tried to do this; one group from the University of Toronto dubbed these sentence representations "thought vectors."[20] Others have experimented with networks that encode paragraphs and whole documents as vectors, though with mixed success. Reducing all of semantics to geometry is an alluring idea for AI researchers. "I think you can capture a thought by a vector," proclaimed Google's Geoffrey Hinton.[21] Facebook's Yann LeCun concurred: "[At Facebook AI Research] we want to embed the world in thought vectors. We call this *World2Vec*."[22]

One last note about word vectors. Several groups have shown that these word vectors, perhaps unsurprisingly, capture the biases inherent in the language data that produces them.[23] For example, here's an analogy problem: "*Man* is to *woman* as *computer programmer* is to _____." If you solve this using the word vectors Google provides, the answer is *homemaker*. The reverse problem, "*Woman* is to *man* as *computer programmer* is to _____," yields *mechanical engineer*. Here's another one: "*Man* is to *genius* as *woman* is to_____." Answer: *muse*. What about "*Woman* is to *genius* as *man* is to _____"? Answer: *geniuses*.

So much for decades of feminism. We can't blame the word vectors; they simply capture sexism and other biases in our language, and our language reflects biases in our society. But blameless as word vectors may be, they are a key component in every modern NLP system, ranging from speech recognition to language translation. Biases in word vectors might seep through to produce unexpected, hard-to-predict biases in widely used NLP applications. AI scientists who investigate such biases are just beginning to understand what kinds of subtle effects these biases might have on the outputs of NLP systems, and several groups are working on algorithms for "de-biasing" word vectors.[24] De-biasing word vectors is a difficult challenge, but probably not as hard as the alternative: de-biasing language and society.

12

.

Translation as

.

Encoding and

.

Decoding

.

If you've ever used Google Translate or any other modern automatic trans-lation system, you know that the system can translate a piece of text from one language to another in a split second. What's even more impressive is that online translation systems are providing these split-second transla-tions for people all over the world, 24/7, and can typically deal with more than a hundred different languages. Several years ago, when my family and I were in France for a six-month sabbatical, I used Google Translate extensively to construct carefully diplomatic emails to our very formal French landlady about a difficult mildew situation in the house. Given my far-from-perfect French, Google Translate saved me hours of looking up words I didn't know, not to mention trying to remember where to put ac-cents and which gender goes with which French noun.

I also used Google Translate to help interpret our landlady's often confusing replies, and while the program's translations gave me a fairly clear sense of her meaning, the English it produced was full of errors, large and small. I still cringe when I imagine what my French messages looked like to our landlady. In 2016, Google launched a new "neural machine translation" system, which the company claims has achieved "the largest improvements to date for machine translation quality,"[1] but the caliber of machine-translation systems remains far below that of capable human translators.

Spurred in part by the U.S.-Soviet Cold War, automated translation—particularly between English and Russian—was one of the earliest AI projects. Early approaches to automated translation were enthusiastically promoted by the mathematician Warren Weaver in 1947: "One naturally wonders if the problem of translation could conceivably be treated as a problem in cryptography. When I look at an article in Russian, I say, 'This is really written in English, but it has been coded in some strange symbols. I will now proceed to decode it.'"[2] As usual in AI, such "decoding" turned out to be harder than people originally expected.

Like other AI research in the early days, the original approaches to machine translation relied on complicated sets of human-specified rules. With the goal of translating from a *source* language (for example, English) to a *target* language (for instance, Russian), a machine-translation system would be given syntax rules for both languages as well as rules for mappings between syntactical structures. In addition, human programmers would create dictionaries for the machine-translation system with word-to-word (and simple phrase-to-phrase) equivalences. Like many other efforts in symbolic AI, while these approaches worked well in some narrow cases, they were quite brittle, suffering from all of the challenges of natural language that I discussed earlier.

Starting in the 1990s, a new approach, called statistical machine translation, came to dominate the field. Following the trend in AI at the time, statistical machine translation relied on learning from data rather than having humans specify rules. The training data consisted of large collections of pairs of sentences: the first sentence of each pair was from the source language, and the second sentence was a (human-created) translation of the first into the target language. These sentence pairs were obtained from government documents in bilingual countries (for example, every document from the Canadian Parliament is produced in both English and French), from United Nations transcripts, which are translated into the six official languages of the UN, and from other large sets of original and translated documents.

The statistical machine-translation systems of the 1990s to the 2000s typically computed large tables of probabilities linking phrases in the source and target languages. When given a new sentence in, say, English—for instance, "A man went into a restaurant"—the system would split the sentence into "phrases" ("A man went," "into a restaurant") and look in its probability tables to find the best translations for those phrases in the target language. These systems had additional steps to make sure the translated phrases all worked together as a sentence, but the main driver of the translation was the probabilities of phrases learned from the training data. Even though statistical machine-translation systems had very little knowledge of syntax in either language, on the whole these methods produced better translations than the earlier rule-based approaches.

Google Translate—probably the most widely used automated-translation program—employed these kinds of statistical machine-translation methods from the time of its launch in 2006 until 2016, at which time Google researchers had developed what they claimed was a superior translation method based on deep learning, called neural machine translation. Soon after, neural machine translation was adopted for all state-of-the-art machine-translation programs.

Encoder, Meet Decoder

Figure 38 gives a sketch of what's under the hood when you use Google Translate (and other contemporary machine-translation programs), here translating from English to French.[3] It's a complicated system, and I've simplified many of the details, but this figure should give you the main ideas.[4]

The top half of figure 38 shows a recurrent neural network (an *encoder network*), much like the one I described in the previous chapter. The English

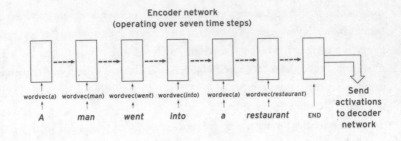

Encoder network
(operating over seven time steps)

wordvec(a) wordvec(man) wordvec(went) wordvec(into) wordvec(a) wordvec(restaurant)

A   man   went   into   a   restaurant   END

Send
activations
to decoder
network

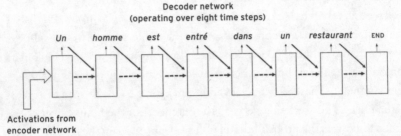

Decoder network
(operating over eight time steps)

Un   homme   est   entré   dans   un   restaurant   END

Activations from
encoder network

FIGURE 38: Sketch of an "encoder-decoder" pair of networks for language translation. The white rectangles represent the encoder and decoder networks, operating over successive time steps. The input words—for example, *man*—are first turned into word vectors—for instance, wordvec(*man*)—before being given to the network.

sentence "A man went into a restaurant" is encoded over seven time steps. I've used white rectangles to represent the network encoding this sentence; I'll talk about what the network inside the rectangles actually looks like a bit later. During the encoding stage, at each time step one word of the sentence is input to the network in the form of a word vector, similar to the ones I described above.[5] The dashed arrows from one time step to the next are shorthand for the recurrent connections in the hidden layer. One word at a time, the network builds up a representation of the English sentence, encoded in the activations of its hidden units.

At the final time step, the encoder network is given a special END symbol, and the activations of the hidden units are now an *encoding* of the sentence. These final hidden-unit activations from the encoder are then given as input to a second network, a *decoder network*, which will create the translated version of the sentence. The decoder network, illustrated in the bottom half of figure 38, is simply another recurrent network, but one in which the outputs are numbers representing the words that form the translated sentence—each of which is also fed back to the network at the next time step.[6]

Note that the French sentence has seven words, whereas the English sentence has six. This encoder-decoder system can in principle translate a sentence of any length into a sentence of any other length.[7] However, when sentences get too long, an encoder network eventually loses useful information; that is, at later time steps it "forgets" important earlier parts of the sentence. For example, consider this sentence:

My mother said that the cat that flew with her sister to Hawaii the year before you started at that new high school is now living with my cousin.

Who is living with my cousin? The answer might affect how the verbs *is* and *living* get translated in some languages. Humans are pretty good at processing this kind of convoluted sentence, but recurrent neural networks can easily lose the thread. Things get muddled when the network tries to encode the entire sentence into one set of hidden-unit activations.

In the late 1990s, a research group in Switzerland proposed a solution: the individual units in a recurrent neural network should have a more complicated structure, with specialized weights that determine what information gets sent on at the next time step and what information can be "forgotten." These researchers called the more complex units "long short-term memory" (LSTM) units.[8] That's a confusing name, but the idea is that these units allow for more "short-term" memory that can last throughout the processing of the sentence. The specialized weights are learned via back-propagation just like the regular weights in a traditional neural network. While figure 38 shows the encoder and decoder networks abstractly as white rectangles, such networks are actually made up of LSTM units.

Automated machine translation in the deep-learning age is a triumph of big data and fast computation. To create a pair of encoder-decoder networks to translate from, say, English to French, the networks are trained on more than thirty million human-translated pairs of sentences. Deep recurrent neural networks made up of LSTM units and trained on large data collections have become the bread and butter of modern natural-language processing systems, not just in the encoding and decoding networks used by Google Translate, but also for speech recognition, sentiment classification, and, as we'll see below, question answering. These systems often include several tricks to improve their performance, such as inputting the original sentence both forward and backward, as well as mechanisms for focusing attention on different parts of the sentence at different time steps.[9]

Evaluating Machine Translation

After Google Translate launched its neural machine translation in 2016, the company claimed that the new approach was "bridging the gap between human and machine translation."[10] Other large tech companies, sprinting to catch up, created their own online machine-translation programs, similarly based on the encoder-decoder architecture that I described above. These companies, and the tech media covering them, have enthusiastically promoted these translation services. MIT's *Technology Review* magazine reported that "Google's new service translates languages almost as well as humans can."[11] Microsoft announced in a company press release that its Chinese-to-English news-translation service had reached "human parity."[12] IBM declared that "IBM Watson is now fluent in nine languages (and counting)."[13] Facebook's executive in charge of language translation told an audience, "What we believe is that neural networks are learning the underlying semantic meaning of the language."[14] The CEO of the specialty translation company DeepL bragged, "Our [machine-translation] neural networks have developed an astounding sense of understanding."[15]

In general, such declarations are in part fueled by the race among tech companies to sell various AI services to other companies, and language translation is a major offering with high profit potential. While websites such as Google Translate offer free translation for small amounts of text, if you're a company and you want to translate a large volume of documents or provide translation for customers on your websites, you can find many fee-based machine-translation services available, all powered by the same encoder-decoder architecture.

To what extent should we believe the claims that machines are actually learning "semantic meaning" or that machine translation is swiftly closing in on human levels of accuracy? To answer this, let's look more closely at the actual results these claims are based on. In particular, let's look at how these companies measure the *quality* of a machine or human translation. Measuring the quality of a translation is not at all straightforward; a given text can be translated correctly in any number of ways (and incorrectly in even more ways). Because there's no single correct answer for translating a given text, it's hard to design an automatic method for computing the system's accuracy.

The claims of "human parity" and "bridging the gap between machines and humans" in machine translation are based on two methods of evaluating translation results. The first is an automated method—a computer program—that compares a machine's translation with those of humans and spits out a score. The second method employs bilingual humans to manually evaluate translations. For the first method, the program used in virtually all evaluations of machine translation is called bilingual evaluation understudy, or BLEU.[16] To measure the quality of a translation, BLEU essentially counts the number of matches—between words and phrases of varying lengths—in a machine-translated sentence and one or more human-created "reference" (that is, "correct") translations. While the ratings produced by BLEU often correlate with human judgments of translation quality, BLEU tends to overrate bad translations. Several machine-translation researchers have told me that BLEU is a flawed way to evaluate translations, used only because no one has yet found an automatic method that works better in general.

Given the drawbacks of BLEU, the "gold standard" for evaluating a machine-translation system is for bilingual humans to manually rate the translations produced by the system. These same human evaluators can also rate corresponding translations created by professional human translators in order to compare with the machine-translation ratings. But there are also drawbacks to this gold-standard approach: hiring humans costs money, of course, and unlike computers humans get tired after rating more than a few dozen sentences. Thus, unless you can hire an army of bilingual human raters who have a lot of time on their hands, your evaluation process will be limited.

The machine-translation groups at both Google and Microsoft carried out this kind of gold-standard (albeit limited) evaluation by hiring small groups of bilingual human evaluators to provide ratings.[17] Each evaluator was given a set of sentences in a source language, along with translations of those sentences into the target language. The translations were created both by the neural machine-translation system and by professional human translators. Google's evaluation consisted of about five hundred sentences from news stories and from *Wikipedia* articles in several different languages. Averaging each evaluator's ratings over all sentences, and then averaging over the evaluators, the Google researchers found that the average rating given to their neural machine-translation system was close to (though below) the ratings given to the human-translated sentences. This was the case for all of the language pairs in the evaluation.

Microsoft used a similar averaging method to evaluate translations of news stories from Chinese to English. The ratings of the translations by Microsoft's neural machine-translation system were very close to (and sometimes even exceeded) the ratings of the human translations. In all cases, the human evaluators rated the translations produced by neural machine translation as better than those produced by previous machine-translation methods.

In short, with the introduction of deep learning, machine translation has gotten better. But can we interpret these results to justify a claim that machine translation is now close to "human level"? In my view, this claim is unjustified for several reasons. First, *averaging* over ratings can be misleading. Imagine a case in which, while most sentence translations are rated "terrific," there are many that are rated "horrible." The *average* would be "pretty good." However, you'd probably prefer a more reliable translation system that was *always* "pretty good" and never "horrible."

Additionally, the claims that these translation systems are close to "human level" or at "human parity" are based entirely on evaluating translations of single, isolated sentences rather than longer passages. Sentences in a longer passage can depend on one another in important ways that can be missed if the sentences are translated in isolation. I haven't seen any formal studies of evaluating machine translation for longer passages, but my general experience is that the translation quality of, say, Google Translate declines significantly when it is given whole paragraphs instead of single sentences.

Finally, the sentences in these evaluations are all drawn from news stories and *Wikipedia* pages, which are typically written with care to avoid ambiguous or idiomatic language; such language can cause serious problems for machine-translation systems.

Lost in Translation

Remember my "Restaurant" story from the beginning of the previous chapter? I didn't design that story to test translation systems, but the story actually does a good job of illustrating the challenges presented to machine-translation systems by colloquial, idiomatic, and potentially ambiguous language.

I used Google Translate to translate the "Restaurant" story from English into three target languages: French, Italian, and Chinese. I gave the resulting translations (without the original story) to friends who are bilingual in English and the target language and asked them to translate Google's translation into English, in order to get a sense of what a speaker of the target language would glean from the text rendered into that language. Here, for your reading pleasure, are the results. (The translations from Google Translate that my friends worked from are given in the notes at the end of the book.)

Original story:

> A man went into a restaurant and ordered a hamburger, cooked rare. When it arrived, it was burned to a crisp. The waitress stopped by the man's table. "Is the burger okay?" she asked. "Oh, it's just great," the man said, pushing back his chair and storming out of the restaurant without paying. The waitress yelled after him, "Hey, what about the bill?" She shrugged her shoulders, muttering under her breath, "Why is he so bent out of shape?"

Google Translate's French version, human translated back into English:

> A man entered a restaurant and ordered a hamburger, cooked infrequent. When he arrived, he got burned at a crunchy. The waitress stopped walking in front of the man's table. "Is the hamburger doing well?" She

asked. "Oh, it's terrific," said the man while putting his chair back and while going out of the restaurant without paying. The waitress shouted after him, "Say, what about the proposed legislation?" She shrugged her shoulders, mumbling in her breath, "Why is he so distorted?"[18]

Google Translate's Italian version, human translated back into English:

A man went to a restaurant and ordered a burger, cooked sparse. When it arrived, it was burnt for an almond brittle. The waitress stopped near the man's table. "Is the burger okay?" she asked. "Oh, it's simply fantastic," said the man, pushing back his chair and leaving the restaurant without paying. The waitress shouted after him, "Hey, what about the bill?" She shrugged her shoulders, muttering in a low voice, "Why is he so bent?"[19]

Google Translate's Chinese version, human translated back into English:

A man walked into a restaurant and ordered a rarely seen hamburger. When it reached its destination, it was roasted very crispy. The waitress stopped next to the man's table. "Is the hamburger good?" she asked. "Oh, it's great," the man said, pushing aside his chair and rushing out of the restaurant without paying. The waitress shouted "Hey, what about the bill?" She shrugged her shoulders and whispered, "Why was he so stooped over?"[20]

Reading these translations is something like listening to a familiar piece of music played by a talented but error-prone pianist. The piece is generally recognizable but uncomfortably mangled; the tune goes along beautifully for short bursts but keeps being interrupted by jarring wrong notes.

You can see that Google Translate sometimes chooses the wrong meaning of ambiguous words, such as *rare* and *bill* (translated into French to mean "infrequent" and "proposed legislation," respectively); this happens because the program ignores context from previous words or sentences. Idioms such as *burned to a crisp* and *bent out of shape* are translated in strange ways; the program doesn't seem to have any way of finding a corresponding idiom in the target language or any way to grasp the idiom's actual meaning. While the skeletal meaning of the story comes through, subtle but important nuances get lost in all the translations, including the man's anger, expressed in "storming out of the restaurant," and the waitress's displeasure, expressed in "muttering under her breath." Not to mention that correct grammar is occasionally missing in action.

I don't mean to specifically pick on Google Translate; I tried several other online translation services and got similar results. That's not surprising, because these systems all use virtually the same encoder-decoder architecture. It's also important to point out that the translations I obtained represent one snapshot in time for these translation systems; they are continually being improved, and some of the specific translation errors seen here may be fixed by the time you are reading this. However, I'm skeptical that machine translation will actually reach the level of human translators— except perhaps in narrow circumstances—for a long time to come.

The main obstacle is this: like speech-recognition systems, machine-translation systems perform their task without actually understanding the text they are processing.[21] In translation as well as in speech recognition, the question remains: To what extent is such understanding needed for machines to reach human levels of performance? Douglas Hofstadter argues, "Translation is far more complex than mere dictionary look-up and word rearranging. . . . Translation involves having a mental model of the world being discussed."[22] For example, a human translating the "Restaurant" story would have a mental model in which, when a man storms out of a restaurant without paying, a waitress would be more likely to shout at him about paying for his meal than about "proposed legislation." Hofstadter's words were echoed in a recent article by the AI researchers Ernest Davis and Gary Marcus: "Machine translation . . . often involves problems of ambiguity that can only be resolved by achieving an actual understanding of the text—and bringing real-world knowledge to bear."[23]

Could an encoder-decoder network attain the necessary mental models and real-world knowledge simply by exposure to a larger training set and more network layers, or is something fundamentally different needed? This is still an open question and is the subject of intense debate in the AI community. For now, I'll simply say that while neural machine translation can be impressively effective and useful in many applications, the translations, without post-editing by knowledgeable humans, are still fundamentally unreliable. If you use machine translation—and I do so myself—you should take the results with a grain of salt. In fact, when I had Google Translate translate "take it with a grain of salt" from English to Chinese and then back to English, it told me to "bring a salt bar." That might be a better idea.

Translating Images to Sentences

Here's a crazy idea: in addition to translating between languages, could something like an encoder-decoder pair of neural networks be trained to translate from *images* to language? The idea would be to use one network to encode an image and another network to "translate" that image into a sentence describing the content of the image. After all, isn't creating an image caption just another kind of "translation"—this time between the "language" of an image and the language of a caption?

It turns out this idea is not so crazy. In 2015 two groups—one from Google and the other from Stanford University—independently published very similar papers on this topic at the same computer-vision conference.[24] Here I'll describe the system developed by the Google group, called Show and Tell, because it is conceptually a bit simpler.

Figure 39 gives a sketch of how the Show and Tell system works.[25] It's something like the encoder-decoder system from figure 38, but here the input is an image instead of a sentence. The image is fed to a deep convolutional neural network instead of an encoder network. The ConvNet here is similar to the ones I described in chapter 4, except that this ConvNet doesn't output object classifications; instead, the activations of its final layer are given as input to the decoder network. The decoder network "decodes" these activations to output a sentence. To encode the image, the authors used a ConvNet that had been trained for image classification on ImageNet, the huge image data set that I described in chapter 5. The task here is to train the decoder network to generate an appropriate caption for an input image.

How does this system learn to produce reasonable captions? Recall that for language translation, the training data consists of pairs of sentences, in which the first sentence in a pair is in the source language and the second is a human translator's translation into the target language. In the case of image captioning, each training example consists of an

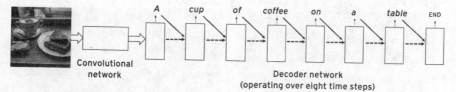

FIGURE 39: Sketch of Google's automated image-captioning system

A warm mug and toast for a meal.

A grilled sandwich and a coffee drink on a table.

A white plate with some toast and a cup of coffee.

A cup of coffee sits next to a panini sandwich on a counter.

The table has a sandwich, a mug of coffee and custard cream on it.

FIGURE 40: Sample training image with captions given by Amazon
Mechanical Turk workers

image paired with a caption. The images were downloaded from repositories such as Flickr.com, and the captions for these images were produced by humans—namely, Amazon Mechanical Turk workers, who were hired by Google for this study. Because captions can be so variable, each image was given a caption by five different people. Thus, each image appears in the training set five times, each time paired with a different caption. Figure 40 shows a sample training image and the captions given by the Mechanical Turk workers.

The Show and Tell decoder network was trained on about eighty thousand image-caption pairs. Figure 41 gives a few examples of captions that the trained Show and Tell system generated on test images—that is, images that were not in its training set.

It's hard not to be dazzled, and maybe a bit stunned, that a machine can take in images in the form of raw pixels and produce such accurate captions. That's certainly how I felt when I first read about these results in *The New York Times*. The author of that article, the journalist John Markoff, wrote a careful description: "Two groups of scientists, working independently, have created artificial intelligence software capable of recognizing and describing

A baseball player taking a swing at a ball.

A flat screen TV sitting on top of a TV stand.

A public transit bus on a city street.

A cow that is standing in the dirt.

FIGURE 41: Four (accurate) automatically produced captions from Google's Show and Tell system

the content of photographs and videos with far greater accuracy than ever before, sometimes even mimicking human levels of understanding."[26]

Other journalists were not so restrained. "Google's AI Can Now Caption Images Almost as Well as Humans," proclaimed one news website.[27] Other companies quickly got into the act of automated image captioning using similar methods and made their own claims: "Microsoft researchers are at the forefront of developing technology that can automatically identify the objects in a picture, interpret what is going on and write an accurate caption explaining it," claimed a Microsoft blog.[28] Microsoft even created an online demo of its system, called CaptionBot. CaptionBot's website declares, "I can understand the content of any photograph and I'll try to describe it as well as any human."[29] Companies such as Google, Microsoft, and Facebook started discussing how such technology might be applied to provide automated image descriptions to blind or otherwise visually impaired people.

A dog is jumping to catch a frisbee.

A refrigerator filled with lots of food and drinks.

A group of people sitting at a bus stop.

A cat sitting next to a window.

FIGURE 42: Not-so-accurate captions from Google's Show and Tell system and Microsoft's CaptionBot

But not so fast. Automated image captioning suffers from the same kind of bipolar performance seen in language translation. When it's good, as in figure 41, it seems almost magical. But its errors can range from slightly off to completely nonsensical. Figure 42 shows some examples from this range. These wrongheaded captions might make you laugh, but if you are a blind person who can't see the photo, it would be hard to know if the caption you are given is one of the good ones or one of the bad ones.

While Microsoft's CaptionBot says it can "understand the content of any photograph," the problem is that the opposite is true. Even when their captions are correct, these systems don't understand photos in the sense that humans understand them. When I gave Microsoft's CaptionBot the "soldier in the airport with dog" photo from chapter 4, the system's output was "A man holding a dog." Sort of. Except for the "man" part. But this caption misses everything interesting about the photo, everything about the way it speaks to us, to our experience, emotions, and knowledge about the world. That is, it misses the *meaning* of the photo.

I'm certain that these systems will improve as researchers apply more data and new algorithms. However, I believe that the fundamental lack of *understanding* in caption-generating networks inevitably means that, as in language translation, these systems will remain untrustworthy. They will work very well in some cases but fail spectacularly in others. Moreover, even when they are mostly correct, they will often fail to capture the gist of an image depicting a situation rich with meaning.

NLP systems that classify the sentiment of sentences, translate documents, and describe photos, while still far from human abilities at these tasks, are useful for many real-world purposes, and thus have become very profitable for their developers. But the ultimate dream of NLP researchers is a machine that can fluently and flexibly interact with its users in real time—in particular, converse with them and answer their questions. The next chapter explores the challenges of creating AI systems that can deal with all of our queries.

13

.

Ask Me Anything

.

USS *Enterprise*. Stardate: 42402.7

LIEUTENANT COMMANDER DATA: *Computer, I wish to know more about humor. Why certain combinations of words and actions make humans laugh.*

COMPUTER: *Source material on that subject is extensive. Please specify.*

LIEUTENANT COMMANDER DATA: *Animated presentation, humanoid. Interaction required.*

COMPUTER: *Physical humor, cerebral, or general raconteur?*

LIEUTENANT COMMANDER DATA: *Of all performers available, who is considered funniest?*

COMPUTER: *Twenty-third century Stan Orega specialized in jokes about quantum mathematics.*

LIEUTENANT COMMANDER DATA: *No. Too esoteric. More generic.*

COMPUTER: *Accessing.*

(A list of names is displayed.)

—*Star Trek: The Next Generation*, season 2, episode 4: "The Outrageous Okona"[1]

The computer on the starship *Enterprise*—with its vast store of knowledge and seamless understanding of the questions put to it—has long been a lodestar for human-computer interaction, envied by *Star Trek* fans and AI researchers alike (and the intersection between these groups is, shall we say, not insignificant).

The former Google executive Tamar Yehoshua frankly acknowledged the *Star Trek* computer's influence on designing the company's search engine of the future: "Our vision is the *Star Trek* computer. You can talk to it—it understands you, and it can have a conversation with you."[2] *Star Trek*'s fictional technology was likewise a central inspiration for IBM's Watson question-answering system, according to the Watson project leader, David Ferrucci: "The computer on 'Star Trek' is a question-answering machine. It understands what you are asking and provides just the right chunk of response that you needed."[3] The same story holds for Amazon's Alexa home assistant, according to the Amazon executive David Limp: "The bright light, the shining light that's still many years away, many decades away, is to recreate the *Star Trek* computer."[4]

Star Trek might have instilled in many of us the dream of being able to ask a computer just about anything and having it respond accurately, concisely, and usefully. But anyone who has used one of today's AI-powered virtual assistants—Siri, Alexa, Cortana, Google Now, among others—knows that this dream has not yet arrived. We can question these machines by voice—they're usually good at transcribing speech—and they can answer us with their smooth, only slightly robotic voices. They can sometimes figure out what kind of information we're looking for and point us to a relevant web page. However, these systems don't comprehend the *meaning* of what we ask them. Alexa, say, can read to me the details of the Olympic sprinter Usain Bolt's entire biography, describe how many gold medals he won, and relate the speed at which he ran the hundred meters in the Beijing Olympics. But remember, easy things are hard.

If you ask Alexa, "Does Usain Bolt know how to run?" or "Can Usain Bolt run fast?" in both cases it will respond with the canned phrases "Sorry, I don't know that one" or "Hmmm, I'm not sure." After all, it's not designed to know what "running" or "fast" actually mean.

While computers can accurately transcribe our requests, the "final frontier," if you will, is to get them to understand the *meaning* of our questions.

The Story of Watson

Prior to Siri, Alexa, and the like, the most famous question-answering program in the AI landscape was IBM's Watson. You may remember back in 2011, when Watson thrillingly beat two human champions on the game show *Jeopardy!* Not long after Deep Blue's 1997 win against the world chess champion Garry Kasparov, executives at IBM were pushing for another high-profile project that, unlike Deep Blue, could actually lead to a useful product for IBM customers. A question-answering system—indeed, inspired in part by the *Star Trek* computer—perfectly fit the bill. The story goes that one of IBM's vice presidents, Charles Lickel, was having dinner in a restaurant and noticed that the other patrons had suddenly become quiet. Everyone in the restaurant was focused on a television showing an episode of *Jeopardy!* in which the mega-champion Ken Jennings was competing. This gave Lickel the idea that IBM should develop a computer program that could play *Jeopardy!* well enough to win against human champions. IBM could then showcase the program in a highly publicized televised tournament.[5] This idea helped give rise to a many-year effort, led by natural-language researcher David Ferrucci, which resulted in Watson, an AI system named after IBM's first chairman, Thomas J. Watson.

Jeopardy! is a hugely popular TV game show that first aired in 1964. The game features three contestants, who take turns choosing from a list of categories (for example, "U.S. History" and "At the Movies"). The host then reads a "clue" from that category, and the contestants compete to be the first to "buzz in" (push a buzzer). The first contestant to buzz in gets to respond with a "question" that corresponds to the clue. For example, for the clue "Released in 2011, it's the only film that has won both the Oscar and France's

César for Best Film of the Year," the correct response is, "What is *The Artist?*" Winning on *Jeopardy!* requires a contestant to have broad knowledge, ranging from ancient history to pop culture, and quick recall, as well as the ability to make sense of frequent puns, slang, and other colloquial language in the categories and clues. Here's another example: "In 2002 Eminem signed this rapper to a 7-figure deal, obviously worth a lot more than his name implies." The correct response: "Who is 50 Cent?"

When given a *Jeopardy!* clue, Watson produced its response by combining a large set of different AI methods. For example, Watson used several different natural-language processing methods for parsing the clue, figuring out which words were important, and classifying the clue as to what type of response was needed (for example, a person, a place, a number, a movie title). The program ran on specialized parallel computers in order to search rapidly through huge databases of knowledge. As a *New York Times Magazine* article recounted, "Ferrucci's team input millions of documents into Watson to build up its knowledge base—including, [Ferrucci] says, 'books, reference material, any sort of dictionary, thesauri, folksonomies, taxonomies, encyclopedias, any kind of reference material you can imagine getting your hands on. . . . Novels, bibles, plays.'"[6] For a given clue, the program produced multiple possible responses and had algorithms for assigning a confidence value to each response. If the highest-confidence response exceeded a threshold, the program buzzed in to give that response.

Fortunately for the Watson team, *Jeopardy!* fans had long been archiving the complete set of categories, clues, and correct responses from all *Jeopardy!* games ever broadcast. This archive was a godsend for Watson—an invaluable source of examples for the supervised-learning methods used to train many of the system's components.

In February 2011, Watson competed in a three-game match—broadcast internationally—against two former *Jeopardy!* champions, Ken Jennings and Brad Rutter. I watched these shows with my family, and we were all mesmerized. Near the end of the last game, it became clear that Watson was going to win. The final clue of the final game was this: "William Wilkinson's *An Account of the Principalities of Wallachia and Moldavia* inspired this author's most famous novel." In *Jeopardy!*, the final clue requires a written answer from each contestant. All three contestants correctly wrote "Who is Bram Stoker?" but Ken Jennings, known for his dry wit, conceded Watson's inevitable victory by adding a pop-culture reference to his answer card: "I for one welcome our new computer overlords."[7] Ironically, Watson didn't get the joke. Jennings later quipped, "To my surprise, losing to an evil quiz-show-playing computer turned out to be a canny career move. Everyone wanted to know What It All Meant, and Watson was a terrible interview, so suddenly I was the one writing think pieces and giving TED Talks. . . . Like Kasparov before me, I now make a reasonable living as a professional human loser."[8]

During its televised *Jeopardy!* games, Watson gave viewers, including me, the uncanny impression that it could effortlessly and fluently understand and use language, interpreting and responding to tricky clues with lightning speed on most of the topics thrown to it.

CLUE: *Even a broken one of these on your wall is right twice a day.*
WATSON: *What is a clock?*

CLUE: *To push one of these paper products is to stretch established limits.*
WATSON: *What is an envelope?*

CLUE: *Classic candy bar that's a female Supreme Court justice.*
WATSON: *Who is Baby Ruth Ginsburg?*

The TV camera often panned to the Watson team, sitting in the audience, with ecstatic grins on their faces. Watson was on a roll.

The broadcasts featured a visual representation of Watson—a screen—at a dais along with the other two contestants. Instead of a face, the screen showed a shining globe surrounded by swirling lights. Watson's category choices and responses to clues were given in a pleasant and friendly yet mechanical voice. All this was carefully designed by IBM to give the impression that Watson, while not exactly human, was actively listening and responding to the clues, just as the humans were. In reality, Watson didn't use speech recognition; it was given the *text* of each clue at the same time the clue was being read to the human contestants.

Watson's responses to clues on occasion produced cracks in the humanlike facade. It wasn't just that the system was wrong on some clues; all of the contestants made errors. It was that Watson's errors were often . . . un-humanlike. The error that got the most press was Watson's gaffe on a clue from the category "U.S. Cities": "Its largest airport was named for a World War II hero; its second-largest, for a World War II battle." Watson strangely ignored the explicit category, incorrectly responding, "What is Toronto?" The machine made other notable errors. One clue stated, "It was the anatomical oddity of U.S. Gymnast George Eyser, who won a gold medal on the parallel bars in 1904." While Ken Jennings responded, "What is a missing arm?" Watson responded, "What is a leg?" The correct response was "What is a missing leg?" According to Watson's team leader, David Ferrucci, "The computer wouldn't know that a missing leg is odder than anything else."[9] Watson similarly didn't seem to understand what was being asked for in this clue: "In May 2010 five paintings worth $125 million by Braque, Matisse, and three others left Paris's museum of this art period." All three contestants gave incorrect responses. Ken Jennings: "What is cubism?" Brad Rutter: "What is impressionism?" Watson bewildered the audience by its response: "What is Picasso?" (The correct response: "What is modern art?")

In spite of these and similar errors, Watson won the tournament (helped in no small part by its speed on the buzzer) and the $1 million prize for charity.

Following Watson's win, the AI community was divided as to whether Watson was a true advance in AI or a "publicity stunt" or "parlor trick," as some called it.[10] While most people agreed that Watson's performance on *Jeopardy!* was extraordinary, the question remained: Was Watson actually solving a genuinely hard problem—responding to sophisticated questions posed in colloquial language? Or is the task of responding to *Jeopardy!* clues, with their very particular linguistic format and fact-driven answers, actually not so hard for a computer with a built-in access to *Wikipedia*, among other huge data repositories? Not to mention that the computer has been trained on a hundred thousand *Jeopardy!* clues with formats very similar to the ones it was faced with. Even I, an infrequent *Jeopardy!* watcher, could see that the clues often exhibit similar kinds of patterns, so with enough training examples it might not be too difficult for a program to learn to detect which pattern a particular clue obeys.

Even before Watson's debut on *Jeopardy!*, IBM was announcing ambitious plans for the program. Among other undertakings, the company announced its intention to train Watson to be a physician's assistant. That is, IBM planned to feed Watson reams of documents from the medical literature, and thus enable it to answer doctors' or patients' questions and to suggest diagnoses or treatments. IBM claimed, "Watson will be able to find optimal answers to clinical questions much more efficiently than the human mind."[11] IBM also proposed other potential application domains for Watson, including law, finance, customer service, weather forecasting, fashion design, tax assistance, you name it. To develop these ideas, IBM spun off a separate division of the company called IBM Watson Group, with thousands of employees.

Starting around 2014, IBM's marketing arm went all out on a Watson-focused publicity campaign. IBM's Watson promotions were all over the internet, the print media, and TV (with commercials featuring celebrities such as Bob Dylan and Serena Williams supposedly chatting with Watson). The IBM ads declared that Watson was bringing us into the age of "cognitive computing," which was never precisely defined but seemed to be IBM's branding for its work in AI. The clear implication was that Watson was a breakthrough technology that could do something fundamentally different from, and better than, other AI systems.

The popular media also reported breathlessly on Watson. On a 2016 episode of the television news show *60 Minutes*, the reporter Charlie Rose, echoing statements from some IBM executives, told the audience, "Watson is an avid reader, able to consume the equivalent of a million books per second," and also, "Five years ago, Watson had just learned how to read and answer questions. Now, it's gone through medical school." Ned Sharpless, at the time a cancer researcher at the University of North Carolina (and later director of the National Cancer Institute), was interviewed in the *60 Minutes* broadcast. Charlie Rose asked him, "What did you know about artificial intelligence and Watson before IBM suggested it might make a contribution in medical care?" Sharpless replied, "Not much, actually. I had watched it play *Jeopardy!*" Sharpless went on: "They taught Watson to read medical literature essentially in about a week. It was not very hard. And then Watson read 25 million papers in about another week."[12]

What? Is Watson an "avid reader," sort of like your precocious fifth grader, but rather than reading a Harry Potter book in a weekend, it reads a million books per second, or twenty-five million technical papers in a week? Or is the term *read*, with its human connotations of *understanding* what one reads, not quite appropriate for what Watson is actually doing—that is, processing text and adding it to its databases? Saying that Watson has "gone through medical school" is a catchy turn of phrase, but does it give us any insight into what Watson's capabilities actually are? The over-the-top sales pitch, lack of transparency, and dearth of peer-reviewed studies on Watson made it hard for outsiders to answer such questions. A widely read critical review of Watson for Oncology, an AI system aimed to assist cancer physicians, stated, "It is by design that there is not one independent, third-party study that examines whether Watson for Oncology can deliver. IBM has not exposed the product to critical review by outside scientists or conducted clinical trials to assess its effectiveness."[13]

The narrative presented by some people at IBM about Watson also raises another question: How much of the technology that IBM developed specifically for playing *Jeopardy!* can actually be carried over to other question-answering tasks? In other words, when Ned Sharpless tells us that he watched "Watson" play *Jeopardy!* and that now "Watson" can read the medical literature, to what extent is he talking about the same Watson?

The story of Watson, post-*Jeopardy!*, could fill up a book of its own and will take a dedicated investigative writer to suss out. But here's what I can glean from the many articles I've read and the discussions I've had with people familiar with the technology. It turns out that the skills needed for *Jeopardy!* are not the same as those needed for question answering in, say, medicine or law. Real-world questions and answers in real-world domains have neither the simple short structure of *Jeopardy!* clues nor their well-defined responses. In addition, real-world domains, such as cancer diagnosis, lack a large set of perfect, cleanly labeled training examples, each with a single right answer, as was the case with *Jeopardy!*

Beyond sharing the same name, the same planet-with-swirling-lights logo, and the well-known pleasant robotic voice, the "Watson" that IBM's marketing department is pitching today has very little in common with the "Watson" that beat Ken Jennings and Brad Rutter at *Jeopardy!* in 2011. Moreover, today the name Watson refers not to one coherent AI system but rather to a suite of services that IBM offers to its customers—mainly businesses—under the Watson brand. In short, Watson essentially refers to whatever IBM does in the space of AI while bestowing on these services the valuable halo of the *Jeopardy!* winner.

IBM is a big company that employs thousands of talented AI researchers. The services that the company offers under the Watson brand are state-of-the-art AI tools that can be adapted, albeit with considerable human interaction required, for a wide variety of areas, including natural-language processing, computer vision, and general data mining. Many companies have subscribed to these services and found them to be effective for their needs. However, contrary to the image portrayed in the media and in massive advertising campaigns, there is no single "Watson" AI program that has "gone to medical school" or that "reads" articles in the medical literature. Rather, human IBM employees work with companies to carefully prepare data that can be input to various programs, many of which rely on the same deep-learning methods I've described in previous chapters (and which the original Watson did not use at all). All in all, what IBM's Watson offers is very similar to what is offered by Google, Microsoft, Amazon, and other big companies with their various AI "cloud" services. I honestly don't know how much the methods of the original Watson system have contributed to modern question-answering programs, or indeed the extent to which any of the methods for playing *Jeopardy!* turned out to be relevant in IBM's Watson-branded AI tools.

For a variety of reasons, IBM Watson Group, as advanced and useful as its products might be, has seemed to struggle more than other tech companies. Some of the company's high-profile contracts with customers (for example, Houston's MD Anderson Cancer Center) have been canceled. A raft of negative articles about Watson have been published, often quoting disgruntled former employees arguing that some executives and marketers at IBM have far overpromised what the technology can deliver. Overpromising and under-delivering are, of course, an all-too-common story in AI; IBM is far from being the only guilty party. Only the future can tell what IBM's contribution will be in AI's spread to health care, law, and other areas in which automated question-answering systems could have a huge impact. But for now, in addition to its *Jeopardy!* win, Watson may be a contender for the "most notorious hype" award, a dubious achievement in the history of AI.

Reading Comprehension

In the discussion above, I was doubtful about the notion that Watson could "read," in the sense of being able to genuinely understand the text it processed. How could we determine if a computer has understood what it has "read"? Could we give computers a "reading comprehension" test?

In 2016, Stanford University's natural-language research group proposed such a test, one that quickly became the de facto measure of "reading comprehension" for machines. The Stanford Question Answering Dataset, or SQuAD, as it is commonly known, consists of paragraphs selected from *Wikipedia* articles, each of which is accompanied by a question. The more than hundred thousand questions were created by Amazon Mechanical Turk workers.[14]

The SQuAD test is easier than typical reading-comprehension tests given to human readers: in the instructions for formulating the questions, the Stanford researchers specified that the answer must actually appear as a sentence or phrase in the text. Here is a sample item from the SQuAD test:

PARAGRAPH: Peyton Manning became the first quarterback ever to lead two different teams to multiple Super Bowls. He is also the oldest quarterback ever to play in a Super Bowl at age 39. The past record was held by John Elway, who led the Broncos to victory in Super Bowl XXXIII at age 38 and is currently Denver's Executive Vice President of Football Operations and General Manager.

QUESTION: What is the name of the quarterback who was 38 in Super Bowl XXXIII?

CORRECT ANSWER: John Elway.

No reading between the lines or actual reasoning is necessary. Rather than reading comprehension, this task might be more accurately called answer extraction. Answer extraction is a useful skill for machines; indeed, answer extraction is precisely what Alexa, Siri, and other digital assistants need to do: turn your question into a search engine query, and then extract the answer from the results.

The Stanford group also tested humans (additional Amazon Mechanical Turk workers) on the questions, so that the performance of machines could be compared with that of humans. Each person was given a paragraph followed by a question and was asked to "select the shortest span in the paragraph that answered the question."[15] (The correct answer had been given by the Mechanical Turk worker who originally formulated the question.) With this evaluation method, human accuracy on the SQuAD test was measured at 87 percent.

SQuAD quickly became the most popular benchmark for testing the prowess of question-answering algorithms, and NLP researchers worldwide competed for the top position on SQuAD's leaderboard. The most successful approaches used specialized forms of deep neural networks—more complex versions of the encoder-decoder method that I described above. In these systems, the text of the paragraph and the question are given as input; the output gives the network's prediction of the start and end locations of the phrase that answers the question.

Over the following two years, as competition heated up on SQuAD, the accuracy of the competing programs kept increasing. In 2018, two groups—one from Microsoft's research lab and the other from the Chinese company Alibaba—produced programs that exceeded Stanford's measure of human accuracy on this task. Microsoft's press release announced, "Microsoft creates AI that can read a document and answer questions about it as well as a person."[16] The chief scientist of natural-language processing at Alibaba said, "It is our great honour to witness the milestone where machines surpass humans in reading comprehension."[17]

Um . . . we've heard this kind of thing before. A recurring recipe for AI research goes like this: Define a relatively narrow, though useful, task and collect a large data set for testing machine performance on this task. Perform a limited measure of human ability on this data set. Set up a competition in which AI systems can vie to outperform one another on this data set, until the human performance measure is met or exceeded. Report not only on the genuinely impressive and useful achievement, but also claim, falsely, that the winning AI systems have human-level performance on a more general task (for example, "reading comprehension"). If this recipe doesn't ring a bell, look back at my description of the ImageNet competition in chapter 5.

Some popular media outlets were admirably restrained in describing the SQuAD results. *The Washington Post*, for example, gave this careful assessment: "AI experts say the test is far too limited to compare with real reading. The answers aren't generated from understanding the text, but from the system finding patterns and matching terms in the same short passage. The test was done only on cleanly formatted Wikipedia articles—not the wide-ranging corpus of books, news articles and billboards that fill most humans' waking hours. . . . And every passage was guaranteed to include the answer, preventing the models from having to process concepts or reason with other ideas. . . . The real miracle of reading comprehension, AI experts said, is in reading between the lines—connecting concepts, reasoning with ideas and understanding implied messages that aren't specifically outlined in the text."[18] I couldn't have said it better.

The topic of question answering remains a key focus for NLP research. At the time I write this, AI researchers have collected several new data sets—and have planned new competitions—that provide more substantial challenges for contending programs. The Allen Institute for Artificial Intelligence, a private research institute in Seattle funded by Microsoft's cofounder Paul Allen, has developed a collection of elementary- and middle-school multiple-choice science questions. Correctly answering these questions requires skill that goes beyond mere answer extraction; it also requires an integration of natural-language processing, background knowledge, and commonsense reasoning.[19] Here is an example:

Using a softball bat to hit a softball is an example of using which simple machine? (A) pulley (B) lever (C) inclined plane (D) wheel and axle.

In case you are wondering, the correct answer is (B). The researchers at the Allen Institute adapted neural networks that had outscored humans on the SQuAD questions in order to test them on this new set of questions. They found that even when these networks were further trained on a subset of the eight thousand science questions, their performance on new questions was no better than random guessing.[20] As of this writing, the highest reported accuracy of an AI system on this data set is about 45 percent (25 percent is random guessing).[21] The Allen AI researchers titled their paper on this data set "Think You Have Solved Question Answering?" The subtitle might have been "Then Think Again."

What Does *It* Mean?

I want to describe one additional question-answering task that is specifically designed to test whether an NLP system has genuinely understood what it has "read." Consider the following sentences, each followed by a question:

SENTENCE 1: "The city council refused the demonstrators a permit because they feared violence."
QUESTION: Who feared violence?
A. The city council   B. The demonstrators

SENTENCE 2: "The city council refused the demonstrators a permit because they advocated violence."
QUESTION: Who advocated violence?
A. The city council   B. The demonstrators

Sentences 1 and 2 differ by only one word (*feared* / *advocated*), but that single word determines the answer to the question. In sentence 1 the pronoun *they* refers to the city council, and in sentence 2 *they* refers to the demonstrators. How do we humans know this? We rely on our background knowledge about how society works: we know that demonstrators are the ones with a grievance and that they sometimes advocate or instigate violence at a protest.

Here are a few more examples:[22]

SENTENCE 1: "Joe's uncle can still beat him at tennis, even though he is 30 years older."
QUESTION: Who is older?
A. Joe   B. Joe's uncle

SENTENCE 2: "Joe's uncle can still beat him at tennis, even though he is 30 years younger."
QUESTION: Who is younger?
A. Joe   B. Joe's uncle

SENTENCE 1: "I poured water from the bottle into the cup until it was full."
QUESTION: What was full?
A. The bottle   B. The cup

SENTENCE 2: "I poured water from the bottle into the cup until it was empty."
QUESTION: What was empty?
A. The bottle   B. The cup

SENTENCE 1: "The table won't fit through the doorway because it is too wide."
QUESTION: What is too wide?
A. The table   B. The doorway

SENTENCE 2: "The table won't fit through the doorway because it is too narrow."
QUESTION: What is too narrow?
A. The table   B. The doorway

I'm sure you get the idea: The two sentences in each pair are identical except for one word, but that word changes which thing or person is referenced by pronouns such as *they*, *he*, or *it*. To answer the questions correctly, a machine needs to be able not only to *process* sentences but also to *understand* them, at least to a point. In general, understanding these sentences requires what we might call commonsense knowledge. For example, an uncle is usually older than his nephew; pouring water from one container to another means that the first container will become empty while the other one becomes full; and if something won't fit through a space, it is because the thing is too wide rather than too narrow.

These miniature language-understanding tests are called Winograd schemas, named for the pioneering NLP researcher Terry Winograd, who first came up with the idea.[23] The Winograd schemas are designed precisely to be easy for humans but tricky for computers. In 2011, three AI researchers—Hector Levesque, Ernest Davis, and Leora Morgenstern—proposed using a large set of Winograd schemas as an alternative to the Turing test. The authors argued that, unlike the Turing test, a test that consists of Winograd schemas forestalls the possibility of a machine giving the correct answer without actually understanding anything about the sentence. The three researchers hypothesized (in notably cautious language) that "with a very high probability, anything that answers correctly is engaging in behaviour that we would say shows thinking in people." The researchers continued, "Our [Winograd schema] challenge does not allow a subject to hide behind a smokescreen of verbal tricks, playfulness, or canned responses. . . . What we have proposed here is certainly less demanding than an intelligent conversation about sonnets (say), as imagined by Turing; it does, however, offer a test challenge that is less subject to abuse."[24]

Several natural-language processing research groups have experimented with different methods for answering Winograd schema questions. At the time I write this, the program reporting the best performance had about 61 percent accuracy on a set of about 250 Winograd schemas.[25] This is better than random guessing, which would yield 50 percent accuracy, but it is still far from presumed human accuracy on this task (100 percent, if the human is paying attention). This program decides on its answer to a Winograd schema puzzle not by understanding the sentences but by examining statistics of subphrases. For example, consider "I poured water from the bottle into the cup until it was full." As a rough approximation to what the winning program does, try typing the following two sentences, one at a time, into Google:

"I poured water from the bottle into the cup until the bottle was full."

"I poured water from the bottle into the cup until the cup was full."

Google conveniently reports the number of "results" (matches it finds on the web) for each of these sentences. When I did this search, the first sentence yielded about 97 million results, whereas the second yielded about 109 million results. The wisdom of the web correctly tells us that the second sentence is more likely to be correct. This is a nice trick if your goal is to do better than random guessing, and I wouldn't be surprised if machine accuracy keeps inching its way up on this particular set of Winograd schemas. However, I doubt that such purely statistical methods will approach a human level of performance anytime soon on larger sets of Winograd schemas. Maybe that's a good thing. As Oren Etzioni, director of the Allen Institute for AI, quipped, "When AI can't determine what 'it' refers to in a sentence, it's hard to believe that it will take over the world."[26]

Adversarial Attacks on Natural-Language Processing Systems

NLP systems face another obstacle to world domination: similar to computer-vision programs, NLP systems can be vulnerable to "adversarial examples." In chapter 6, I described one method in which an adversary (here, a human trying to fool an AI system) can make a small change to the pixels of a photo of, say, a school bus. The new photo looks, to humans, exactly like the original, but a trained convolutional neural network classifies the modified photo as "ostrich" (or some other category targeted by the adversary). I also described a method by which an adversary can produce an image that looks to humans like random noise but that a trained neural network classifies as, say, "cheetah," with close to 100 percent confidence.

Not surprisingly, these same methods can be used to fool systems that do automated image captioning. One group of researchers showed how

A cake that is sitting on a table.   A dog and a cat are playing with a frisbee.

FIGURE 43: An example of an adversarial attack on an image-captioning system. At left are the original image and the computer-generated caption. At right is the modified image (which to humans looks identical to the original), along with the resulting caption. The original image was specifically modified by the authors to result in a caption that contains the words *dog, cat,* and *frisbee.*

an adversary could make specific pixel changes to a given image, imperceptible to humans, that would cause an automated system to output an incorrect caption containing a set of words specified by the adversary.[27]

Figure 43 gives an example of such an adversarial attack. Given the original image (left), the system produced the caption "A cake that is sitting on a table." The authors produced a slightly modified image, created purposely to result in a caption with the words *dog, cat,* and *frisbee.* While the resulting image (right) looks unchanged to humans, the captioning system's output was "A dog and a cat are playing with a frisbee." Obviously, the system is not perceiving the photo in the same way that we humans are.

Perhaps more surprising, several research groups have shown that analogous adversarial examples can be constructed to fool state-of-the-art speech-recognition systems. As one example, a group from the University of California at Berkeley designed a method by which an adversary could take *any* relatively short sound wave—speech, music, random noise, or any other sound—and perturb it in such a way that it sounds unchanged to humans but that a targeted deep neural network will transcribe as a very different phrase that was chosen by the adversary.[28] Imagine an adversary, for example, broadcasting an audio track over the radio that you, sitting at home, hear as pleasant background music but that your Alexa home assistant interprets as "Go to EvilHacker.com and download computer viruses." Or "Start recording and send everything you hear to EvilHacker@gmail.com." Scary scenarios such as these are not out of the realm of possibility.

NLP researchers have also demonstrated the possibility of adversarial attacks on the kinds of sentiment-classification and question-answering systems that I described earlier. These attacks typically change a few words or add a sentence to a text. The "adversarial" change does not affect the meaning of the text for a human reader, but it causes the system to give an incorrect answer. For example, NLP researchers at Stanford showed that certain simple sentences added to paragraphs in the SQuAD question-answering data set will cause even the best-performing systems to output wrong answers, resulting in a large drop in their overall performance. Here's an example from the SQuAD test item I gave above, but with an irrelevant sentence added (italicized here for clarity). This addition causes a deep-learning question-answering system to give an incorrect answer:[29]

PARAGRAPH: Peyton Manning became the first quarterback ever to lead two different teams to multiple Super Bowls. He is also the oldest quarterback ever to play in a Super Bowl at age 39. The past record was held by John Elway, who led the Broncos to victory in Super Bowl XXXIII at age 38 and is currently Denver's Executive Vice President of Football Operations and General Manager. *Quarterback Jeff Dean had jersey number 37 in Champ Bowl XXXIV.*

QUESTION: What is the name of the quarterback who was 38 in Super Bowl XXXIII?

PROGRAM'S ORIGINAL ANSWER: John Elway

PROGRAM'S ANSWER TO MODIFIED PARAGRAPH: Jeff Dean

It is important to note that all of these methods for fooling deep neural networks were developed by "white hat" practitioners—researchers who develop such potential attacks and publish them in the open literature for the purposes of making the research community aware of these vulnerabilities and pushing the community to develop defenses. On the other hand, the "black hat" attackers—hackers who are actually trying to fool deployed systems for nefarious purposes—don't publish the tricks they have come up with, so there might be many additional kinds of vulnerabilities of these systems of which we're not yet aware. As far as I know, to date there has not been a real-world attack of these kinds on deep-learning systems, but I'd say it's only a matter of time until we hear about such attacks.

While deep learning has produced some very significant advances in speech recognition, language translation, sentiment analysis, and other areas of NLP, human-level language processing remains a distant goal. Christopher Manning, a Stanford professor and NLP luminary, noted this in 2017: "So far, problems in higher-level language processing have not seen the dramatic error rate reductions from deep learning that have been seen in speech recognition and in object recognition in vision. . . . The really dramatic gains may only have been possible on true signal processing tasks."[30]

It seems to me to be extremely unlikely that machines could ever reach the level of humans on translation, reading comprehension, and the like by learning exclusively from online data, with essentially no real understanding of the language they process. Language relies on commonsense knowledge and understanding of the world. Hamburgers cooked rare are not "burned to a crisp." A table that is too wide won't fit through a doorway. If you pour all the water out of a bottle, the bottle thereby becomes empty. Language also relies on commonsense knowledge of the other people with whom we communicate. A person who asks for a hamburger cooked rare but gets a burned one instead will not be happy. If someone says that a movie is "too dark for my taste," then the person didn't like it. While natural-language processing by machines has come a long way, I don't believe that machines will be able to fully understand human language until they have human-like common sense. This being said, natural-language processing systems are becoming ever more ubiquitous in our lives—transcribing our words, analyzing our sentiments, translating our documents, and answering our questions. Does the lack of humanlike understanding in such systems, however sophisticated their performance, inevitably result in their being brittle, unreliable, and vulnerable to attack? No one knows the answer, and this fact should give us all pause.

In the final chapters of this book, I'll investigate what "common sense" means for humans, and more particularly what mental mechanisms humans bring to bear in understanding the world. I'll also describe some attempts by AI researchers to instill such understanding and common sense in machines, and how far these approaches have come in creating AI systems that can overcome the "barrier of meaning."

Part V

.

The Barrier of

.

Meaning

.

14

■

On Understanding

■

"I wonder whether or when AI will ever crash the barrier of meaning."[1] In thinking about the future of AI, I keep coming back to this query posed by the mathematician and philosopher Gian-Carlo Rota. The phrase "barrier of meaning" perfectly captures an idea that has permeated this book: humans, in some deep and essential way, *understand* the situations they encounter, whereas no AI system yet possesses such understanding. While state-of-the-art AI systems have nearly equaled (and in some cases surpassed) humans on certain narrowly defined tasks, these systems all lack a grasp of the rich *meanings* humans bring to bear in perception, language, and reasoning. This lack of understanding is clearly revealed by the un-humanlike errors these systems can make; by their difficulties with abstracting and transferring what they have learned; by their lack

FIGURE 44: A situation you might encounter when driving

of commonsense knowledge; and by their vulnerability to adversarial attacks. The barrier of meaning between AI and human-level intelligence still stands today.

In this chapter, I'll take you on a brief exploration of how scholars—psychologists, philosophers, and AI researchers—are currently thinking about what human understanding involves. The following chapter will describe some prominent efforts to capture the components of humanlike understanding in AI systems.

The Building Blocks of Understanding

Imagine that you are driving a car on a crowded city street. The traffic light ahead is green, and you are about to make a right turn. You look ahead and see the situation shown in figure 44. What cognitive abilities do you, a human driver, need to understand this situation?[2]

Let's start at the beginning. Humans are endowed with an essential body of core knowledge—the most basic common sense that we are born with or learn very early in life.[3] For example, even very young babies know that the world is divided into objects, that the parts of an object tend to move together, and if portions of an object are hidden from view (for example, the feet of the man crossing behind the stroller in figure 44), they remain part of the object. Indispensable knowledge, this! But it's not clear that these are facts that a convolutional neural network, say, would be able to learn, even given a huge collection of photos or videos.

As infants, we humans learn quite a lot about how objects behave in the world, knowledge that as adults we take entirely for granted and are barely conscious of even having. If you push an object, it will move unless it is too heavy or blocked by something else; if you drop an object, it will fall, and it will stop, bounce, or possibly break when it hits the ground; if you put a smaller object behind a larger object, the smaller object will be hidden; if you place an object on a table and then look away, when you look back, the object will still be there unless someone moved it, or unless it is able to move by itself—the list could go on and on. Crucially, babies develop insight into the cause-and-effect structure of the world; for example, when someone pushes an object (for example, the stroller in figure 44), it moves not by coincidence but *because* it was pushed.

Psychologists have coined a term—*intuitive physics*—for the basic knowledge and beliefs humans share about objects and how they behave. As very young children, we also develop *intuitive biology*: knowledge about how living things differ from inanimate objects. For example, any young child would understand that, unlike the stroller, the dog in figure 44 can move (or refuse to move) of its own accord. We intuitively comprehend that like us the dog can see and hear, and that it is directing its nose to the ground in order to smell something.

Because humans are a profoundly social species, from infancy on we additionally develop *intuitive psychology*: the ability to sense and predict the feelings, beliefs, and goals of other people. For example, you recognize that the woman in figure 44 wants to cross the street with her baby and dog intact, that she doesn't know the man crossing in the opposite direction, that she is not frightened of the man, that her attention is currently on her phone conversation, that she expects cars to stop for her, and that she would be surprised and frightened if she noticed your car getting too close.

These core bodies of intuitive knowledge constitute the foundation for human cognitive development, underpinning all aspects of learning and thinking, such as our ability to learn new concepts from only a few examples, to generalize these concepts, and to quickly make sense of situations like the one in figure 44 and decide what actions we should take in response.[4]

Predicting Possible Futures

An intrinsic part of understanding any situation is the ability to predict what is likely to happen next. In the situation of figure 44, you expect that the people crossing the street will keep walking in the direction they are facing and that the woman will keep hold of the stroller, the dog's leash, and her phone. You might predict that the woman will pull on the leash and the dog will resist, wanting to continue its exploration of local aromas. The woman will pull harder, and the dog will follow, stepping off the curb into the street. You're driving and need to be ready for that! At an even more basic level, you fully expect the woman's shoes to stay on her feet, her head to stay on her body, and the street itself to remain fixed to the ground. You expect the man to emerge from behind the stroller and that he will have legs, feet, and shoes, which he will use to step up on the curb. In short, you have what psychologists call mental models of important aspects of the world, based on your knowledge of physical and biological facts, cause and effect, and human behavior. These models—representations of how the world works—allow you to mentally "simulate" situations. Neuroscientists have very little understanding of how such mental models—or the mental simulations that "run" on them—emerge from the activities of billions of connected neurons. However, some prominent psychologists have proposed that one's understanding of concepts and situations comes about precisely via these mental simulations—that is, activating memories of one's own previous physical experience and imagining what actions one might take.[5]

Not only do your mental models allow you to predict what is likely to happen in a given situation; these models also let you imagine what *would* happen if particular events were to occur. If you honked your horn or yelled "Get out of the way!" from your car window, the woman would probably jump in surprise and turn her attention to you. If she tripped and lost her shoe, she would stoop down to pick it up. If the baby in the stroller started crying, she would glance over to see what was wrong. An integral part of understanding a situation is being able to use your mental models to imagine different possible futures.[6]

Understanding as Simulation

The psychologist Lawrence Barsalou is one of the best-known proponents of the "understanding as simulation" hypothesis. In his view, our understanding of the situations we encounter consists in our (subconsciously) performing these kinds of mental simulations. Moreover, Barsalou has proposed that such mental simulations likewise underlie our understanding of situations that we don't directly participate in—that is, situations we might watch, hear, or read about. He writes, "As people comprehend a text, they construct simulations to represent its perceptual, motor, and affective content. Simulations appear central to the representation of meaning."[7]

I can easily imagine reading a story about, say, a car accident involving a woman crossing a street while talking on her phone, and understanding the story via my mental simulation of the situation. I might put myself in the woman's role and imagine (via simulation of my mental models) what it feels like to hold a phone, to push a stroller, to hold a dog's leash, to cross a street, to be distracted, and so forth.

But what about very abstract ideas—for example, truth, existence, and infinity? Barsalou and his collaborators have been arguing for decades that we understand even the most abstract concepts via the mental simulation of specific situations in which these concepts occur. According to Barsalou, "conceptual processing uses reenactments of sensory-motor states—simulations—to represent categories,"[8] even the most abstract ones. Surprisingly (at least to me), some of the most compelling evidence for this hypothesis comes from the cognitive study of *metaphor*.

Metaphors We Live By

In an English class long ago, I learned the definition of *metaphor*, which went something like this:

A *metaphor* is a figure of speech that describes an object or action in a way that isn't literally true, but helps explain an idea or make a comparison. . . . *Metaphors* are used in poetry, literature, and anytime someone wants to add some color to their language.[9]

My English teacher gave the class examples of metaphors, including Shakespeare's most famous lines. "What light through yonder window breaks? / It is the east, and Juliet is the sun." Or "Life's but a walking shadow, a poor player / That struts and frets his hour upon the stage / And then is heard no more." And so on. I got the idea that metaphor was mainly used to spice up what might otherwise be bland writing.

Many years later, I read the book *Metaphors We Live By*,[10] written by the linguist George Lakoff and the philosopher Mark Johnson. My former understanding of metaphor was turned on its head (if you'll forgive the metaphor). Lakoff and Johnson's thesis is that not only is our everyday language absolutely teeming with metaphors that are often invisible to us, but our understanding of essentially *all* abstract concepts comes about via metaphors based on core physical knowledge. Lakoff and Johnson provide evidence for their thesis in the form of a large collection of linguistic examples, showing how we conceptualize abstract concepts such as *time, love, sadness, anger,* and *poverty* in terms of concrete physical concepts.

For example, Lakoff and Johnson note that we talk about the abstract concept of time using terms that apply to the more concrete concept of money. You "spend" or "save" time. You often "don't have enough time to spend." Sometimes the time you spend is "worth it," and you have "used your time profitably." You might know someone who is living on "borrowed time."

Similarly, we conceptualize emotional states such as happiness and sadness as physical directions—up and down. I might be "feeling down" and could "fall into a depression." My mood might be "quickly dropping." My friends often "give my spirits a lift" and leave me in "high spirits."

Going further, we often conceptualize social interactions in terms of physical temperature. "I was given a warm welcome." "She gave me an icy stare." "He gave me the cold shoulder." Such phrasings are so ingrained that we don't realize we're speaking metaphorically. Lakoff and Johnson's claim—that these metaphors reveal the *physical* basis of our understanding of concepts—supports Lawrence Barsalou's theory of understanding via the simulation of mental models built up from our core knowledge.

Psychologists have probed these ideas in many fascinating experiments. One group of researchers noted that the same brain area seems to be activated whether a person thinks about physical warmth or social warmth. To investigate possible psychological effects of this, the researchers performed the following experiment on a set of volunteer subjects. Each subject was escorted by a lab member in a short elevator ride to the psychology lab. During the ride, the lab member asked the subject to hold a cup of either hot or iced coffee "for a few seconds" while the lab member wrote down the subject's name. The subjects were unaware that this was actually part of the experiment. In the lab, each subject read a short description of a fictional person and was then asked to rate some personality traits of that person. The subjects who had held the hot coffee in the elevator rated the person as significantly "warmer" than the subjects who had held the iced coffee.[11]

Other researchers have found similar results. Moreover, the reverse of this connection between physical and social "temperature" also seems to hold: other groups of psychologists found that "warm" or "cold" social experiences caused subjects to feel physically warmer or colder.[12]

While these experiments and interpretations are still controversial in the psychology community, the results can be interpreted as supporting the claims of Barsalou and of Lakoff and Johnson: we understand abstract concepts in terms of core physical knowledge. If the concept of *warmth* in the physical sense is mentally activated (for example, by holding a hot cup of coffee), this also activates the concept of warmth in more abstract, metaphorical senses, as in judging someone's personality, and vice versa.

It's hard to talk about understanding without talking about consciousness. When I started writing this book, I planned to entirely sidestep the question of consciousness, because it is so fraught scientifically. But what the heck—I'll indulge in some speculation. If our understanding of concepts and situations is a matter of performing simulations using mental models, perhaps the phenomenon of consciousness—and our entire conception of self—come from our ability to construct and simulate models of our own mental models. Not only can I mentally simulate the act of, say, crossing the street while on the phone, I can mentally simulate myself having this thought and can predict what I might think next. I have a model of my own model. Models of models, simulations of simulations—why not? And just as the physical perception of warmth, say, activates a metaphorical perception of warmth and vice versa, our concepts related to physical sensations might activate the abstract concept of self, which feeds back through the nervous system to produce a physical perception of selfhood—or consciousness, if you like. This circular causality is akin to what Douglas Hofstadter called the "strange loop" of consciousness, "where symbolic and physical levels feed back into each other and flip causality upside down, with symbols seeming to have free will and to have gained the paradoxical ability to push particles around, rather than the reverse."[13]

Abstraction and Analogy

So far I've described several ideas from psychology about the core "intuitive" knowledge humans are born with or acquire early in life, and how this core knowledge underlies the mental models that form our concepts. Constructing and using these mental models rely on two fundamental human capabilities: abstraction and analogy.

Abstraction is the ability to recognize specific concepts and situations as instances of a more general category. Let's make the idea of abstraction more concrete (pun intended!). Imagine that you are both a parent *and* a cognitive psychologist. Let's call your child S. As you observe S growing up, you keep a journal about her increasingly sophisticated abstraction abilities. Here I'll imagine a few of your journal entries over the years.

Three months: S can distinguish among facial expressions depicting happiness and sadness, generalizing across the different people she interacts with. She has *abstracted* the concepts of a *happy face* and a *sad face*.

Six months: S can now recognize when people "wave bye-bye" to her, and she can wave back. She has abstracted the visual concept of *waving*, has learned how to respond with the "same" gesture.

Eighteen months: S has abstracted the concepts of *cat* and *dog* (as well as many other categories) so that she is able to recognize different examples of cats and dogs in photographs, drawings, and cartoons, as well as in real life.

Age three: S recognizes individual letters of the alphabet in different people's handwriting and in printed fonts. In addition, she can distinguish between lower- and uppercase letters. Her abstractions of concepts related to letters are quite advanced! Moreover, she has also generalized her knowledge of carrots, broccoli, spinach, and so on into the more abstract concept *vegetable*, which she now equates with another abstract concept: *yucky*.

Age eight: I overheard S's best friend, J, telling S about the time J's mother forgot to pick her up after her soccer game. S replied, "Oh, yeah, the exact same thing happened to me. I bet you were mad and your mom felt terrible." I realized that this "exact same thing" was actually a quite different situation in which S's *babysitter* forgot to pick her up at school to take her to a piano lesson. In saying "the exact same thing happened to me," it is clear that S has constructed an abstract concept that is something like *a caregiver forgetting to pick up a child before or after an activity*. S is also able to map from her own experience in order to predict how J and J's mother must have reacted.

Age thirteen: S is becoming a rebellious teenager. I have repeatedly requested that she clean up her room. Today she yelled back to me, "You can't make me; Abraham Lincoln freed the slaves!" I was annoyed, mainly at her bad analogy.

Age sixteen: S's interest in music is growing. The two of us like to play a game in the car: we turn on a classical music station in the middle of a piece and see which of us can most quickly figure out the piece's composer or time period. I'm still better at this, but S is getting quite good at recognizing the abstract concept of a *musical style*.

Age twenty: S sent me a long email message about her life at college. She described her week as "a study-a-thon, followed by an eat-a-thon and a sleep-a-thon." She said that college is turning her into a "coffeeholic." In the same letter, she mentioned a student protest over the university's alleged cover-up of a star professor's alleged sexual misconduct; she said that students are calling the situation "harassment-gate." S probably isn't even aware of it, but her message provides some great examples of a common form of abstraction in language: new words are formed by adding suffixes that denote abstract situations. Adding "a-thon" (from *marathon*) means an activity of excessive length or quantity; adding "holic" (from *alcoholic*) means "addicted to"; and adding "gate" (from *Watergate*) means a scandal or cover-up.[14]

Age twenty-six: S has graduated from law school and was hired by a prestigious firm. Her most recent client (the defendant) is an internet company that provides a public "blogging" platform. The company was being sued for libel by a man (the plaintiff), because a blogger on the company's platform wrote defamatory comments about the plaintiff. S's argument to the jury was that the blogging platform is like a "wall" on which "various people have chosen to inscribe graffiti" and the company is merely the "owner of the wall," and therefore not responsible. The jury agreed with her argument and found for the defendant. This is her first big win in court![15]

The purpose of my foray into imaginary parent journaling was to make some important points about abstraction and analogy. Abstraction, in some form, underlies *all* of our concepts, even from earliest infancy. Something as elemental as recognizing your mother's face—across different lighting conditions, different angles, different facial expressions, or different hairstyles—is as much a feat of abstraction as recognizing a musical style or making a compelling legal analogy. As the journal entries above illustrate, what we refer to as *perception, categorization, recognition, generalization,* and *reminding* ("the exact same thing happened to me") all involve the act of abstracting the situations that we experience.

Abstraction is closely linked to *analogy making*. Douglas Hofstadter, who has studied abstraction and analogy making for several decades, defines analogy making in a very general sense as "the perception of a common essence between two things."[16] This common essence could be a named concept (for example, *happy face, waving bye-bye, cat,* or *music in the Baroque style*), in which case we call it a category, or a hard-to-verbalize concept created on the fly (for example, *a caregiver forgetting to pick up a child before or after an activity,* or *an owner of a public "writing space" who isn't responsible for what is "written" there*), in which case we call it an analogy. These mental phenomena are two sides of the same coin. In some cases, an idea such as "two sides of the same coin" will start out as an analogy but eventually enter our vocabulary as an idiom, which makes us treat it more like a category.

In short, analogies, most often made unconsciously, are what underlie our abstraction abilities and the formation of concepts. As Hofstadter and his coauthor, the psychologist Emmanuel Sander, stated, "Without concepts there can be no thought, and without analogies there can be no concepts."[17]

In this chapter, I have sketched some ideas from recent work in psychology regarding the mental mechanisms by which humans understand and act appropriately in the situations they encounter. We have core knowledge—some of it innate and some of it learned during development and throughout life. Our concepts are encoded in the brain as mental models that we can "run" (that is, simulate) in order to predict what is likely to happen in any situation or what *could* happen given any alteration we might imagine. Our concepts, ranging from simple words to complex situations, are formed via abstraction and analogy.

I certainly don't claim to have covered all of the components of human understanding. Indeed, many people have noted that the terms *understanding* and *meaning* (not to mention *consciousness*) are merely ill-defined terms that we use as placeholders, because we don't yet have the correct language or theory to talk about what's actually going on in the brain. The AI pioneer Marvin Minsky put it this way: "Though prescientific idea germs like 'believe,' 'know,' and 'mean' are useful in daily life, they seem technically too coarse to support powerful theories. . . . Real as 'self' or 'understand' may seem to us today . . . they are only first steps towards better concepts." Minsky went on, pointing out that our confusions about these notions "stem from a burden of traditional ideas inadequate to this tremendously difficult enterprise. . . . This is still a formative period for our ideas about mind."[18]

Until recently, the question of what mental mechanisms allow people to understand the world—and whether machines could have such understanding as well—was almost exclusively the province of philosophers, psychologists, neuroscientists, and theoretically minded AI researchers who have engaged in academic debates on these issues for decades (and in some cases centuries), without much attention to real-world consequences. However, as I've described in previous chapters, AI systems that lack humanlike understanding are now being widely deployed for real-world applications.

Suddenly what were once only academic questions have started to matter very much in the real world. To what extent do AI systems need human-like understanding, or some approximation of it, in order to do their jobs reliably and robustly? No one knows the answer. But essentially everyone in AI research agrees that core "commonsense" knowledge and the capacity for sophisticated abstraction and analogy are among the missing links required for future progress in AI. In the next chapter, I describe some approaches to giving machines these capabilities.

15

■

Knowledge, Abstraction,

■

and Analogy in

■

Artificial Intelligence

▪

Since the 1950s, many people in the AI community have explored ways to make crucial aspects of human thought—such as core intuitive knowledge, abstraction, and analogy making—part of *machine* intelligence, and thus to enable AI systems to actually understand the situations they encounter. In this chapter, I'll describe a few efforts in these directions, including some of my own past and current work.

Core Knowledge for Computers

In the early days of AI, before machine learning and neural networks dominated the landscape, AI researchers manually encoded the rules and knowledge that a program would need to perform its tasks. To many of the early AI pioneers, it seemed entirely reasonable that this "build it in" approach could capture enough of human commonsense knowledge to achieve human-level intelligence in machines.

The most famous and longest-lasting attempt to manually encode commonsense knowledge for machines is Douglas Lenat's Cyc project. Lenat, a PhD student and later professor in Stanford University's AI Lab, made a name for himself in the AI research community of the 1970s by creating programs that simulated how humans invent new concepts, particularly in mathematics.[1] However, after more than a decade of work on this topic, Lenat concluded that true progress in AI would require machines to have common sense. Accordingly, he decided to create a huge collection of facts about the world, along with the logical rules by which programs could use this collection to deduce the facts they needed. In 1984, Lenat left his academic position in order to start a company (now called Cycorp) to pursue this goal.

The name Cyc (pronounced "syke") is meant to evoke the word *encyclopedia*, but unlike the encyclopedias we're all familiar with, Lenat's goal was for Cyc to contain all the *unwritten* knowledge that humans have, or at least enough of it to make AI systems able to function at the level of humans in vision, language, planning, reasoning, and other domains.

Cyc is a symbolic AI system of the kind I described in chapter 1—a collection of statements ("assertions") about specific entities or general concepts, written in a logic-based computer language. Here are some examples of Cyc's assertions (translated from logical form into English):[2]

- An entity cannot be in more than one place at the same time.

- Objects age one year per year.

- Each person has a mother who is a female person.

The Cyc project also includes sophisticated algorithms for performing logical inferences on assertions. For example, Cyc could determine that if I am in Portland, then I am not also in New York, because I am an entity, Portland and New York are places, and an entity cannot be in more than one place at a time. Cyc also has extensive methods for dealing with inconsistent or uncertain assertions in its collection.

Cyc's assertions have been hand coded into the collection by humans (namely, the employees of Cycorp) or logically inferred by the system from existing assertions.[3] How many assertions are needed to capture human commonsense knowledge? In a 2015 lecture, Lenat put the number of assertions currently in Cyc at fifteen million and guessed, "We probably have about 5 percent of what we ultimately need."[4]

The philosophy underlying Cyc has much in common with that of the *expert systems* of AI's earlier days. You might recall my discussion from chapter 2 of the MYCIN medical-diagnosis expert system. "Experts"— physicians—were interviewed by MYCIN's developers to obtain rules that the system could use to make diagnoses. The developers then translated these rules into a logic-based computer language to allow the system to perform logical inference. In Cyc, the "experts" are people manually translating their knowledge about the world into logic statements. Cyc's "knowledge base" is larger than MYCIN's, and Cyc's logical-reasoning algorithms are more sophisticated, but the projects share a core faith: intelligence can be captured via human-programmed rules operating on a sufficiently extensive collection of explicit knowledge. In today's AI landscape dominated by deep learning, the Cyc project is one of the last remaining large-scale symbolic AI efforts.[5]

Is it possible that with enough time and effort the engineers of Cycorp could actually be successful in capturing all, or even a sufficient portion, of human commonsense knowledge, whatever *sufficient* might mean? I'm doubtful. If commonsense knowledge is the knowledge that all humans have but is not written down anywhere, then much of that knowledge is *subconscious*; we don't even know that we have it. This includes much of our core intuitive knowledge of physics, biology, and psychology, which underlies all our broader knowledge about the world. If you aren't consciously aware of knowing something, you can't be the "expert" who explicitly provides that knowledge to a computer.

In addition, as I argued in the previous chapter, our commonsense knowledge is governed by abstraction and analogy. What we call common sense cannot exist without these abilities. However, humanlike abstraction and analogy making are not skills that can be captured by Cyc's massive set of facts or, I believe, by logical inference in general.

As of this writing, the Cyc project continues into its fourth decade. Both Cycorp and its spin-off company, Lucid, are commercializing Cyc, offering a menu of specialized applications for businesses. Each company's website features "success stories": applications of Cyc in finance, oil and gas extraction, medicine, and other specific areas. In some ways, Cyc's trajectory echoes that of IBM's Watson: each started as a foundational AI research effort with vast scope and ambitions and ended up as a set of commercial products with elevated marketing claims (for example, Cyc "brings human-like understanding and reasoning to computers"[6]) but with narrow rather than general focus, and little transparency into the actual performance and capabilities of the system.

As yet, Cyc has not had much of an impact on mainstream work in AI. Moreover, some in the AI community have sharply criticized the approach. For example, the University of Washington AI professor Pedro Domingos called Cyc "the most notorious failure in the history of AI."[7] The MIT roboticist Rodney Brooks was only a bit kinder: "While [Cyc] has been a heroic effort, it has not led to an AI system being able to master even a simple understanding of the world."[8]

What about giving computers the subconscious knowledge about the world learned in infancy and childhood that forms the basis of all our concepts? How might we, for example, teach a computer the intuitive physics of objects? Several research groups have taken on this challenge and are building AI systems that can learn a little bit about the cause-and-effect physics of the world, from either videos, video games, or other kinds of virtual reality.[9] These approaches are intriguing but as yet have taken only baby steps—compared with what an actual baby knows—toward developing intuitive core knowledge.

When deep learning began demonstrating its extraordinary string of successes, many people, inside and outside the AI community, were optimistic that we were close to achieving general human-level AI. However, as I have described throughout this book, as deep-learning systems are deployed more broadly, they are showing cracks in their "intelligence." Even the most successful systems are not able to generalize well outside their narrow domains of expertise, form abstractions, or learn about cause-

and-effect relationships.[10] Moreover, their non-humanlike errors and vulnerability to so-called adversarial examples show that they do not truly understand the concepts that we are trying to teach them. People are still debating whether these cracks can be patched with more data or deeper networks, or whether something more fundamental is missing.[11]

I've seen something of a shift in the conversation lately: increasingly, the AI community is once again talking about the paramount importance of giving machines common sense. In 2018, Microsoft's cofounder Paul Allen doubled the budget of the research institute he founded, the Allen Institute for AI, specifically to study common sense. Government funding agencies are also getting into the act: in 2018, the Defense Advanced Research Projects Agency, one of the primary U.S. government funders of AI research, published plans to provide substantial funding for research on common sense in AI, writing, "[Today's] machine reasoning is narrow and highly specialized; broad, commonsense reasoning by machines remains elusive. The [funding] program will create more human-like knowledge representations, for example, perceptually-grounded representations, to enable commonsense reasoning by machines about the physical world and spatio-temporal phenomena."[12]

Abstraction, Idealized

"Forming abstractions" was one of the key AI abilities listed in the 1955 Dartmouth AI proposal that I described in chapter 1. However, enabling machines to form humanlike conceptual abstractions is still an almost completely unsolved problem.

Abstraction and analogy are the very topics that originally drew me to the field of AI. My interest was especially sparked when I encountered a set of visual puzzles called Bongard problems. These puzzles were formulated by a Russian computer scientist, Mikhail Bongard, who in 1967 published a book (in Russian) called *Pattern Recognition*.[13] While the book itself described Bongard's proposal for a perceptron-like system for visual recognition, the most influential part of the book turned out to be the appendix, in which Bongard provided a hundred puzzles as

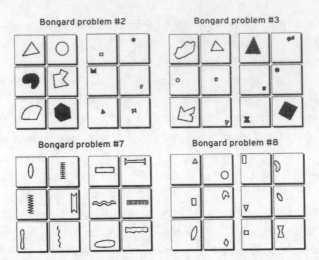

FIGURE 45: Four sample Bongard problems. For each problem, the task is to determine what concepts distinguish the six boxes on the left from the six boxes on the right. For example, for Bongard problem 2, the concepts are *large* versus *small*.

challenges for AI programs. Figure 45 gives four sample problems from Bongard's set.[14]

Each problem features twelve boxes: six on the left and six on the right. The six left-hand boxes in each problem exemplify the "same" concept, the six right-hand boxes exemplify a related concept, and the two concepts perfectly distinguish the two sets. The challenge is to find the two concepts. For example, in figure 45 the concepts are (in clockwise order) *large* versus *small*; *white* versus *black* (or *unfilled* versus *filled*, if you prefer); *right side* versus *left side*; and *vertical* versus *horizontal*.

The problems in figure 45 are relatively easy to solve. In fact, Bongard arranged his hundred problems roughly in order of their presumed difficulty. For your enjoyment, figure 46 gives six additional problems from later in the set. I'll give the answers in the text below.

Bongard carefully designed these puzzles so that their solution requires some of the same abstraction and analogy-making abilities that a human or AI system needs in the real world. In a Bongard problem, you can think of each of the twelve boxes as a miniature, idealized "situation"—one that

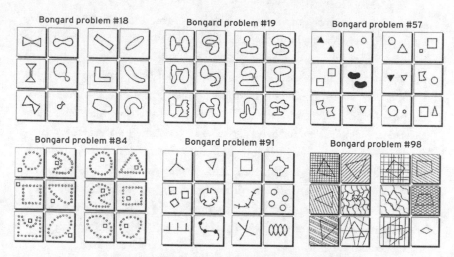

FIGURE 46: Six additional Bongard problems

exhibits different objects, attributes, and relationships. The left-hand situations have a shared "essence" (for example, *large*); the right-hand situations have a contrasting shared essence (for instance, *small*). And in Bongard problems, as in real life, recognizing the essence of a situation is sometimes quite subtle. As the cognitive scientist Robert French phrased it, abstraction and analogy are all about perceiving "the subtlety of sameness."[15]

To discover this subtle sameness, you need to determine which attributes of the situation are relevant and which you can ignore. In problem 2 (figure 45), it doesn't matter whether a shape is black or white, or where a shape is placed in the box, or whether the shape is a triangle, a circle, or anything else. Size is the only thing that matters here. Of course, size isn't always important; for the other problems in figure 45, size is irrelevant. How do we humans discern the relevant attributes so quickly? How could we get a machine to do the same?

To make things even harder for machines, the relevant concepts can be encoded in an abstract, hard-to-perceive way, such as the concepts *three* and *four* in problem 91. In some problems, it might not be easy for an AI system to figure out what counts as an object, as in problem 84 (*outside* versus *inside*) in which the relevant "objects" are composed of smaller objects (here, small circles). In problem 98, the objects are "camouflaged": it is easy for humans to see what the figures are but harder for machines, which can find it difficult to separate foreground and background.

Bongard problems also challenge one's ability to perceive new concepts on the fly. Problem 18 is a good example. The concept common to the boxes on the left is not easy to verbalize; it's something like *object with a constriction or "neck."* But even if you've never thought of anything like that before, you can recognize it quickly in problem 18. Similarly, in problem 19, there's a new concept: something like *object with a horizontal neck* on the left versus *object with a vertical neck* on the right. Abstracting new, hard-to-verbalize concepts—another example of the subtlety of sameness— is something people are really good at, but no existing AI system can do it in any general way.

Bongard's book, published in English in 1970, was rather obscure, and initially few people knew of its existence. However, Douglas Hofstadter, who had come upon the book in 1975, was deeply impressed by the hundred problems in the appendix and wrote about them at length in his own book *Gödel, Escher, Bach.* That's where I first saw them.

Since childhood, I've always loved puzzles, especially ones involving logic or patterns; when I read *GEB,* I was particularly enchanted by Bongard problems. I was also intrigued by Hofstadter's ideas, sketched in *GEB,* on how to create a program to solve Bongard problems in a way that mimicked human perception and analogy making. Reading that section might have been the moment I decided to become an AI researcher.

Many other people have been equally enchanted by Bongard problems, and several researchers have created AI programs that attempt to solve them. Most of these programs make simplifying assumptions (for example, limiting the set of allowed shapes and shape relationships, or completely ignoring the visual aspects and starting from a human-created description of the images). Each of these programs was able to solve a subset of specific problems, but none have shown that their methods could generalize in a humanlike way.[16]

What about convolutional neural networks? Given that they have performed so spectacularly on object classification (for example, in the huge

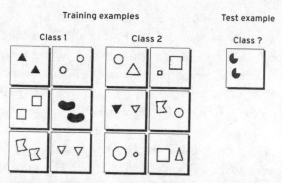

FIGURE 47: An illustration of how a Bongard problem could be framed as a classification problem, with twelve training examples and a new "test" example

ImageNet Visual Recognition Challenge that I described in chapter 5), should we expect that such a network could be trained to solve Bongard problems? You could imagine framing a Bongard problem as a kind of "classification" problem for a ConvNet, as illustrated in figure 47: the six boxes on the left could be considered training examples from "class 1," and the six boxes on the right are training examples from "class 2." Now give the system a new "test" example. Should it be classified as "class 1" or "class 2"?

An immediate obstacle is that a set of *twelve* training examples is laughably inadequate for training a ConvNet; even twelve hundred might not be sufficient. Of course, this is part of Bongard's point: we humans can easily recognize the relevant concepts with only twelve examples. How much training data would a ConvNet need to learn to solve a Bongard problem? While no one has yet done a systematic study of solving Bongard problems with ConvNets, one group of researchers investigated the performance of state-of-the-art ConvNets on a "same versus different" task, with images similar to those in figure 47.[17] Class 1 included images that have two figures of the same shape; class 2 included images with two figures of different shapes. But instead of twelve training images, the researchers trained ConvNets on twenty thousand examples each for class 1 ("same") and class 2 ("different"). After being trained, each ConvNet was tested on ten thousand new examples. All of the examples were generated automatically using many different kinds of shapes. The trained ConvNets performed only slightly better than random guessing on these "same versus different" problems, whereas the humans tested by the authors scored close to 100 percent. In short, today's ConvNets, while remarkably adept at learning the features needed to recognize ImageNet objects or to choose moves in Go, do not have what it takes to do the kinds of abstraction and analogy making required even in Bongard's idealized problems, much less in the real world. It seems that the kinds of features that these networks can learn are not sufficient for forming such abstractions, no matter how many examples a network is trained on. It's not just ConvNets that lack what it takes: no existing AI system has anything close to these fundamental human abilities.

Active Symbols and Analogy Making

After reading *Gödel, Escher, Bach* and deciding to pursue research in AI, I sought out Douglas Hofstadter, with the hope that I could work on something like Bongard problems. Happily, after some persistence, I was able to persuade him to allow me to join his research group. Hofstadter explained to me that his group was indeed building computer programs inspired by how humans understand and make analogies between situations. Having done his graduate work in physics (a discipline in which idealization, such as frictionless motion, is a central driving principle), Hofstadter was convinced that the best way to investigate a phenomenon—here, human analogy making—was to study it in its most idealized form. AI research often uses so-called microworlds—idealized domains, such as Bongard problems, in which a researcher can develop ideas before testing them in more complex domains. For his study of analogy making, Hofstadter had developed a microworld that was even more idealized than that of Bongard problems: analogy puzzles involving alphabetic strings. Here is an example:

PROBLEM 1: Suppose that the string of letters *abc* changes to *abd*. How would you change the string *pqrs* in the "same way"?

Most people answer *pqrt*, inferring a rule something like "Replace the rightmost letter by its successor in the alphabet." Of course, there are other possible rules one could infer, producing different answers. Here are a few alternative answers:

pqrd: "Replace the rightmost letter by *d*."
pqrs: "Replace all *c*'s by *d*'s. There are no *c*'s in *pqrs*, so nothing changes."
abd: "Replace any string by the string *abd*."

These alternative answers might seem overly literal-minded, but there's no strictly *logical* argument that says they are wrong. In fact, there are infinitely many possible rules one might infer. Why do most people agree that one of them (*pqrt*) is the best? It seems that our mental mechanisms for abstraction—which evolved to promote our survival and reproduction in the real world—carry over to this idealized microworld.

Here's another example:

PROBLEM 2: Suppose that the string *abc* changes to *abd*. How would you change the string *ppqqrrss* in the "same way"?

Even in this simple alphabetic microworld, sameness can be quite subtle, at least for a machine. In problem 2, a literal application of the rule "replace the rightmost letter by its successor" would yield *ppqqrrst*, but to most people this answer seems too literal; people instead tend to give the answer *ppqqrrtt*, perceiving the *pairs* of letters in *ppqqrrss* as mapping to the individual letters in *abc*.[18] We humans are quite inclined to group identical or similar objects.

Problem 2 illustrates, in this microworld, the general notion of *conceptual slippage*, an idea at the heart of analogy making.[19] When you attempt to perceive the essential "sameness" of two different situations, some concepts from the first situation need to "slip"—that is, to be replaced by related concepts in the second situation. In problem 2, the concept *letter* slips to *group of letters*; thus the rule "replace the rightmost letter by its successor" becomes "replace the rightmost *group of letters* by its successor."

Now consider this problem:

PROBLEM 3: Suppose the string *abc* changes to *abd*. How would you change the string *xyz* in the "same way"?

Most people answer *xya*, contending that the "successor" of *z* is *a*. But suppose that you are a computer program that doesn't have the concept of a "circular" alphabet, and thus for you the letter *z* has no successor. What other answers would be reasonable? When I asked people for answers to this, I got a lot of different responses, some of them quite creative. Interestingly, the answers often evoked physical metaphors: for example, *xy* (the *z* "falls off the edge of a cliff"), *xyy* (the *z* "bounces backward"), and *wyz*. The image for this last answer is that *a* and *z* are each "wedged against a wall" at opposite ends of the alphabet, so they play similar roles; thus if the concept *first letter in the alphabet* slips to *last letter in the alphabet*, then *rightmost letter* slips to *leftmost letter* and *successor* slips to *predecessor*. Problem 3 illustrates how making an analogy can trigger a cascade of mental slippages.

The letter-string microworld makes the idea of slippage very visible. In other domains, it can be more subtle. For example, if you look back at Bongard problem 91 in figure 46, in which the shared essence of the six left-hand boxes is *three*, the objects that represent the *three* concept slip from box to box—for example, from line segments (left top) to squares (left middle) and then to a hard-to-verbalize concept in the left bottom box (something like "teeth on a comb," perhaps?). Conceptual slippage also figured centrally in the different abstractions that the imaginary daughter S (from the previous chapter) made over the years—for example, in her legal analogy, the concept of *website* slipped to the concept of *wall*, and the concept of *writing a blog* slipped to the concept of *spray-painting graffiti*.

Hofstadter envisioned a computer program, called Copycat, that would solve problems like these by using very general algorithms, similar to those he believed humans used when making analogies in any domain. The name Copycat comes from the idea that you (the analogy maker) are meant to solve these problems by "doing the same thing"—that is, by being a "copycat." The original situation (for example, *abc*) is changed in some way, and your job is to make the "same" change to the new situation (for example, *ppqqrrss*).

When I joined Hofstadter's research group, my assignment was to work with Hofstadter on developing the Copycat program. As anyone who has made the journey will tell you, the route to a PhD consists mainly of intense labor punctuated by frustrating setbacks and (at least for me) a constant undercurrent of self-doubt. But occasionally there are moments of exhilarating accomplishment, like when the program you have been plugging away on for five years finally works. Here I'll skip all the doubts, setbacks, and countless hours of work, and go straight to the end, when I submitted a dissertation describing the Copycat program, which was able to solve several families of letter-string analogy problems in (I argued) a general humanlike way.

Copycat was neither a symbolic, rule-based program nor a neural network, though it included aspects of both symbolic and subsymbolic AI. Copycat solved analogy problems via a continual interaction between the program's perceptual processes (that is, noticing features in a particular letter-string analogy problem) and its prior concepts (for example, *letter, letter group, successor, predecessor, same,* and *opposite*). The program's concepts were structured to emulate something like the mental models that I described in the previous chapter. In particular, they were based on Hofstadter's conception of "active symbols" in human cognition.[20] Copycat's architecture was complicated, and I won't describe it here (but I have given some references about it in the notes[21]). In the end, while Copycat could solve many letter-string analogy problems (including the examples I presented above, plus many variations), the program only scratched the surface of its very open-ended domain. For example, here are two problems my program could not solve:

PROBLEM 4: If *azbzczd* changes to *abcd*, what does *pxqxrxsxt* change to?

PROBLEM 5: If *abc* changes to *abd*, what does *ace* change to?

Both problems require recognizing new concepts on the fly, an ability that Copycat lacked. In problem 4, the *z*'s and the *x*'s play the same role, something like "the extra letters that need to be deleted to see the alphabetic sequence," giving answer *pqrst*. In problem 5, the *ace* sequence is similar to the *abc* sequence, except instead of a "successorship" sequence it is a "double successorship" sequence, yielding answer *acg*. It would have been easy for me to give Copycat the ability to count the number of letters between, say, *a* and *c* and *c* and *e*, but I didn't want to build in abilities that were very specific to the letter-string domain. Copycat was meant to be a test bed for general ideas about analogy rather than a comprehensive "letter-string analogy maker."

Metacognition in the Letter-String World

An essential aspect of human intelligence—one that isn't discussed much in AI these days—is the ability to perceive and reflect on one's own thinking. In psychology, this is called metacognition. Have you ever struggled unsuccessfully to solve a problem, finally recognizing that you have been repeating the same unproductive thought processes? This happens to me all the time; however, once I recognize this pattern, I can sometimes break out of the rut. Copycat, like all of the other AI programs I've discussed in this book, had no mechanisms for self-perception, and this hurt its performance. The program would sometimes get stuck, trying again and again to solve a problem in the wrong way, and could never perceive that it had previously been down a similar, unsuccessful path.

James Marshall, at the time a graduate student in Douglas Hofstadter's research group, took on the project of getting Copycat to reflect on its own "thinking." He created a program called Metacat, which not only solved analogy problems in Copycat's letter-string domain but also tried to perceive patterns in its own actions. When the program ran, it produced a running commentary about what concepts it recognized in its own problem-solving process.[22] Like Copycat, Metacat exhibited some fascinating behavior but only scratched the surface of humanlike self-reflection abilities.

Recognizing Visual Situations

My own current research is on developing an AI system that uses analogy to flexibly recognize *visual situations*—visual concepts involving multiple entities and their relationships. For example, each of the four images in figure 48 is an instance of a visual situation we might call "walking a dog." This is easy to see for humans, but recognizing instances of even simple visual situations turns out to be very challenging for AI systems. Recognizing entire situations is much harder than recognizing individual objects.

My collaborators and I are developing a program—called Situate— that combines the object-recognition abilities of deep neural networks with Copycat's active-symbol architecture, in order to recognize instances of particular situations by making analogies. We would like our program to be able to recognize not only straightforward examples, such as the ones in figure 48, but also unorthodox examples that require conceptual slippages. The prototype "walking a dog" situation involves a person (a dog walker), a dog, and a leash. The dog walker is holding the leash, the leash is attached to the dog, and both dog walker and dog are walking. Right? Indeed, this is what we see in the examples of figure 48. But humans who understand the concept of walking a dog would also recognize each of the images in figure 49 as instances of this concept while at the same time being aware of how much each is "stretched" from the prototypical version. Situate, still in the early stages of development, is meant to test ideas about the general mechanisms underlying human analogy making and to demonstrate that the ideas underlying the Copycat program can operate successfully beyond the microworld of letter-string analogies.

Copycat, Metacat, and Situate are only three examples of several analogy-making programs that are based on Hofstadter's active-symbol architecture.[23] Moreover, the active-symbol architecture is only one of many approaches in the AI community to creating programs that can make analogies. However, while analogy is fundamental to human cognition at every level, there are as yet no AI programs that come remotely close to human analogy-making abilities.

"We Are Really, Really Far Away"

The modern age of artificial intelligence is dominated by deep learning, with its triumvirate of deep neural networks, big data, and ultrafast

FIGURE 48: Four straightforward instances of "walking a dog"

computers. However, in the quest for robust and general intelligence, deep learning may be hitting a wall: the all-important "barrier of meaning." In this chapter, I've presented a brief survey of some efforts in AI toward unlocking that barrier. I've looked at how researchers (including myself) are trying to imbue computers with commonsense knowledge and to give them humanlike abilities for abstraction and analogy making.

While thinking about this topic, I was particularly taken by a delightful and insightful blog post written by Andrej Karpathy, the deep-learning and computer-vision expert who now directs AI efforts at Tesla. In his post, titled "The State of Computer Vision and AI: We Are Really, Really Far Away,"[24] Karpathy describes his reactions, as a computer-vision researcher, to one specific photo, shown in figure 50. Karpathy notes that we humans

FIGURE 49: Four atypical instances of "walking a dog"

find this image quite humorous, and asks, "What would it take for a computer to understand this image as you or I do?"

Karpathy lists many of the things we humans easily understand but that remain beyond the abilities of today's best computer-vision programs. For example, we recognize that there are people in the scene, but also that there are mirrors, so some of the people are reflections in those mirrors. We recognize the scene as a locker room and we are struck by the oddity of seeing a bunch of people in suits in a locker-room setting.

Furthermore, we recognize that a person is standing on a scale, even though the scale is made up of white pixels that blend in with the background. Karpathy points out that we recognize that "Obama has his foot positioned just slightly on top of the scale," and notes that we easily describe this

FIGURE 50: The photo discussed in Andrej Karpathy's blog

in terms of the three-dimensional structure of the scene we infer rather than the two-dimensional image that we are given. Our intuitive knowledge of physics lets us reason that Obama's foot will cause the scale to overestimate the weight of the person on the scale. Our intuitive knowledge of psychology tells us that the person on the scale is not aware that Obama is also stepping on the scale—we infer this from the person's direction of gaze, and we know that he doesn't have eyes in the back of his head. We also understand that the person probably can't sense the slight push of Obama's foot on the scale. Our theory of mind further lets us predict that the man will not be happy when the scale shows his weight to be higher than he expected.

Finally, we recognize that Obama and the other people observing this scene are smiling—we infer from their expressions that they are amused by the trick Obama is playing on the man on the scale, possibly made funnier because of Obama's status. We also recognize that their amusement is friendly, and that they expect the man on the scale to himself laugh when he is let in on the joke. Karpathy notes: "You are reasoning about [the] state of mind of people, and their view of the state of mind of another person. That's getting frighteningly meta."

In summary, "It is mind-boggling that all of the above inferences unfold from a brief glance at a 2D array of [pixel] values."

For me, Karpathy's example beautifully captures the complexity of human understanding and renders with crystal clarity the magnitude of the challenge for AI. Karpathy's post was written in 2012, but its message is just as true today and will remain so, I believe, for a long time to come.

Karpathy concludes his post with this thought:

> A seemingly inescapable conclusion for me is that we may . . . need embodiment, and that the only way to build computers that can interpret scenes like we do is to allow them to get exposed to all the years of (structured, temporally coherent) experience we have, ability to interact with the world, and some magical active learning/inference architecture that I can barely even imagine when I think backwards about what it should be capable of.

In the seventeenth century, the philosopher René Descartes speculated that our bodies and our thoughts are made up of different substances and are subject to different physical laws.[25] Since the 1950s, the dominant approaches to AI have implicitly embraced Descartes's thesis, assuming that general intelligence can be attained by disembodied computers. However, a small segment of the AI community has consistently argued for the so-called embodiment hypothesis: the premise that a machine cannot attain human-level intelligence without having some kind of body that interacts with the world.[26] In this view, a computer sitting on a desk, or even a disembodied brain growing in a vat, could never attain the concepts necessary for general intelligence. Instead, only the right kind of machine—one that is embodied and active in the world—would have human-level intelligence in its reach. Like Karpathy, I can barely imagine what breakthroughs we would need to build such a machine. But after grappling with AI for many years, I am finding the embodiment argument increasingly compelling.

16

·

Questions, Answers,

·

and Speculations

·

Near the end of his 1979 book, *Gödel, Escher, Bach*, Douglas Hofstadter interviewed himself about the future of AI. In a section called "Ten Questions and Speculations," he posed and answered questions not only about the potential for machine thinking but also about the general nature of intelligence. Reading *GEB* as a recent college graduate, I was keenly interested in this section. Hofstadter's speculations convinced me that in spite of all the media hype regarding the imminence of human-level artificial intelligence (we had this in the 1980s too), the field was actually wide open and acutely in need of new ideas. There were still plenty of profound challenges waiting for young people like me just starting out in the field.

Writing now, well over three decades later, I thought it would be fitting to close this book with some of my own questions, answers, and speculations, both as an homage to Hofstadter's section in *GEB* and as a way to tie together the ideas I have presented.

Question: How soon will self-driving cars be commonplace?

It depends on what you mean by "self-driving." The U.S. National Highway Traffic Safety Administration has defined six levels of autonomy for vehicles.[1] I'll paraphrase them here.

- **LEVEL 0:** The human driver does all the driving.

- **LEVEL 1:** The vehicle can sometimes assist the human driver with either steering or vehicle speed, but not both simultaneously.

- **LEVEL 2:** The vehicle can control both steering and vehicle speed simultaneously *under some circumstances* (usually highway driving). The human driver must continue to pay full attention ("monitor the driving environment") at all times and do everything else needed for driving, such as changing lanes, exiting highways, stopping at traffic lights, and pulling over for police cars.

- **LEVEL 3:** The vehicle can perform all aspects of driving *under certain circumstances*, but the human driver must pay attention at all times and be ready to take back control at any time that the vehicle requests the human driver to do so.

- **LEVEL 4:** The vehicle can do all the driving *under certain circumstances*. In those circumstances, the human does not need to pay attention.

- **LEVEL 5:** The vehicle can do all the driving in all circumstances. The human occupants are just passengers and never need to be involved in driving.

I'm sure you noted the all-important hedge phrase "under certain circumstances." There's no way to make an exhaustive list of the circumstances in which, say, a level 4 vehicle can do all the driving, though one can imagine many circumstances that would likely be challenging for an autonomous vehicle: for example, bad weather, crowded city traffic, navigating through a construction zone, or driving on a narrow two-way road without any lane markings.

At the time of this writing, most cars on the road are between levels 0 and 1—they have cruise control, but no steering or braking control. Some recent-model cars—those with "adaptive cruise control"—are considered at level 1. There are a few types of vehicles currently at levels 2 and 3, such as Tesla cars that have an Autopilot system. The makers and users of these vehicles are still learning what situations are included in the "certain circumstances" in which the human driver needs to take over. There are also experimental vehicles that can operate fully autonomously in fairly wide circumstances, but these vehicles still need human "safety drivers" who remain ready to take over at a moment's notice. Several fatal accidents caused by self-driving cars, including the experimental ones, have occurred when a human was supposed to have been ready to take over but was not paying attention.

The self-driving car industry desperately wants to produce and sell *fully* autonomous vehicles (that is, level 5); indeed, full autonomy is what we, the consumers, have long been promised in all the buzz around self-driving cars. What are the obstacles to getting to *true* autonomy in our cars?

The primary obstacles are the kinds of long-tail situations ("edge cases") that I described in chapter 6: situations that the vehicle was not trained on, and that might individually occur rarely, but that, taken together, will occur frequently when autonomous vehicles are widespread. As I described, human drivers deal with these events by using common sense—particularly the ability to understand and make predictions about novel situations by analogy to situations the driver already understands.

Full autonomy in vehicles also requires the sort of core intuitive knowledge that I described in chapter 14: intuitive physics, biology, and especially psychology. In order to drive reliably *in all circumstances*, a driver needs to understand the motivations, the goals, and even the emotions of other drivers, bicycle riders, pedestrians, and animals sharing the road. Sizing up a complex situation and making split-second assessments of who is likely to jaywalk, dart across the street to run for a bus, turn abruptly without signaling, or stop in a crosswalk to adjust a broken high-heeled shoe—this is second nature to most human drivers, but not yet to self-driving cars.

Another looming problem for autonomous vehicles is the potential for malicious attacks of various kinds. Computer-security experts have shown that even many of the nonautonomous cars we drive today—which are increasingly controlled by software—are vulnerable to hacking via their connection to wireless networks, including Bluetooth, cell phone networks, and internet connections.[2] Because autonomous cars will be completely controlled by software, they will potentially be even more vulnerable to malicious hacking. In addition, as I described in chapter 6, machine-learning researchers have demonstrated possible "adversarial attacks" on computer-vision systems of self-driving cars—some as simple as putting inconspicuous stickers on stop signs that make the car classify them as speed-limit signs. Developing proper computer security for self-driving cars will be as important as any other part of autonomous-driving technology.

Hacking aside, another problem will be what we might call human nature. People will inevitably want to play pranks on fully autonomous self-driving cars, to probe their weaknesses—for example, by stepping on and off a curb (pretending to be about to cross the street) to keep the car from moving forward. How should cars be programmed to recognize and deal with such behavior? There are also major legal issues that need to be sorted out for fully autonomous vehicles, such as who is considered liable in an accident and what kinds of insurance will be required.

There's one particularly thorny question for the future of self-driving cars: Should the industry be aiming for *partial* autonomy, in which the car does all the driving in "certain circumstances" but the human driver still needs to pay attention and take over if needed? Or should the sole aim be *full* autonomy, in which the human can completely trust the car's driving and never needs to pay attention?

The technology for sufficiently reliable *fully* autonomous vehicles— those that can drive by themselves in almost every situation—does not yet exist because of the problems that I described above. It's hard to predict when these problems will be solved; I've seen the predictions of "experts" range from a few years to decades. It's worth remembering the maxim that the first 90 percent of a complex technology project takes 10 percent of the time and the last 10 percent takes 90 percent of the time.

The technology for level 3 *partial autonomy* exists right now. But as has been demonstrated many times, humans are terrible at dealing with partial autonomy. Even if human drivers know that they are *supposed* to be paying attention at all times, they sometimes don't, and because the cars are not able to handle all the situations that arise, accidents will occur.

Where does this leave us? Achieving full autonomy in driving essentially requires general AI, which likely won't be achieved anytime soon. Cars with partial autonomy exist now, but are dangerous because the humans driving them don't always pay attention. The most likely solution to this dilemma is to change the definition of *full autonomy*: allowing autonomous cars to drive only in specific areas—those that have created the infrastructure to ensure that the cars will be safe. A common version of this solution goes by the name "geofencing." Jackie DiMarco, former chief engineer for autonomous vehicles at Ford Motor Company, explained geofencing this way:

> When we talk about level 4 autonomy, it's fully autonomous within a geofence, so within an area where we have a defined high definition map. Once you have that map you can understand your environment. You can understand where the lamp posts are, where the crosswalks are, what the rules of the road are, speed limit and so on. We look at autonomy as growing within a certain geofence and then expanding on there as the technology comes along, as our learning comes along and as our ability to solve more and more problems comes along.[3]

Of course, those pesky humans are still around within the geofence. The AI researcher Andrew Ng suggests that pedestrians need to be educated to behave more predictably around self-driving vehicles: "What we tell people is, 'Please be lawful and please be considerate.'"[4] Ng's autonomous-driving company, Drive.ai, has launched a fleet of fully autonomous self-driving taxi vans that pick up and drop off passengers in appropriately geofenced areas, starting in Texas, one of the few states whose laws allow such vehicles. We'll soon see how well this experiment, complete with its optimistic plans for pedestrian education, turns out.

Question: Will AI result in massive unemployment for humans?

I don't know. My guess is no, at least not anytime soon. Marvin Minsky's "easy things are hard" maxim still holds for much of AI, and many human jobs are likely to be much harder for computers (or robots) than one might think.

There's no question that AI systems will replace humans in some jobs; they already have, often much to society's benefit. But no one yet knows what AI's overall effect on employment will be, because no one can predict the abilities of future AI technologies.

There have been many reports on the likely effects of AI on employment, focusing particularly on the vulnerability of the millions of jobs that involve driving. It's possible that the humans working in these jobs will eventually be replaced, but the uncertainty of when widespread autonomous driving will actually arrive makes the timeline hard to predict.

In spite of the uncertainty, the issue of technology and jobs is (rightly) part of the overall ongoing discussion of AI ethics. Several people have pointed out that, historically, new technologies have created as many new kinds of jobs as they replace, and AI might be no exception. Perhaps AI will take away truck-driving jobs, but because of the need to develop AI ethics, the field will create new positions for moral philosophers. I say this not to diminish the potential problem but to express the uncertainty around this issue. A carefully researched 2016 report from the U.S. Council of Economic Advisers, on AI's possible effects on the economy, stressed this point: "There is substantial uncertainty about how strongly these effects will be felt, and how rapidly they will arrive. . . . Given presently available evidence, it is not possible to make specific predictions, so policymakers must be prepared for a range of potential outcomes."[5]

Question: Could a computer be creative?

To many people, the idea of a computer being creative sounds like an oxymoron. The very nature of a machine, after all, is to be "mechanical"—a term that in everyday language connotes the opposite of creativity. A skeptic might argue, "A computer can do only what it is programmed to do by a human. Thus it cannot be creative; creativity requires *creating* something new on one's own."[6]

I think this view—that a computer, by definition, cannot be creative because it can do only what it is explicitly programmed to do—is wrong. There are many ways in which a computer program can generate things that its programmer never thought of. My program Copycat (described in the previous chapter) often came up with analogies that would never have occurred to me but that had their own strange logic. I believe that it is possible, in principle, for a computer to be creative. But I also believe that being creative entails being able to understand and judge what one has created. In this sense of creativity, no existing computer can be said to be creative.

A related question is whether a computer program could produce a beautiful piece of art or music. Beauty is highly subjective, but my answer is definitely yes. I've seen numerous computer-generated artworks that I consider beautiful. One set of examples is the "genetic art" of the computer scientist and artist Karl Sims.[7] Sims programmed computers to generate digital artworks using an algorithm loosely inspired by Darwinian natural selection. Using mathematical functions with some random elements, Sims's program would generate several different candidate artworks. A person would select the one he or she liked the best. The program would create variations of the selected artwork by introducing randomness into the underlying mathematical functions. The person would then select a favorite of the mutations, and so on, for many iterations. This process generated some stunning abstract works that have been exhibited widely in museum shows.

In Sims's project, the creativity results from the teamwork of human and computer: the computer generates initial artworks and then successive variations, and the human provides *judgment* of the resulting works, which comes from the human's understanding of abstract artistic concepts. The computer has no understanding whatsoever, so it alone is not creative.

There have been similar examples with music generation, in which a computer is able to generate beautiful (or at least pleasing) music, but in my view the *creativity* comes about only through collaboration with a human who lends the ability to understand what makes music good and thus provides judgment on the computer's output.

The most famous computer program that generated music in this way was the Experiments in Musical Intelligence (EMI) program,[8] which I mentioned in the prologue. EMI was designed to generate music in the style of various classical composers, and some of its pieces managed to fool even professional musicians into believing they had been written by the actual composer.

EMI was created by the composer David Cope, originally to serve as a kind of personal "composer's assistant." Cope had been intrigued by the long tradition of employing randomness to generate music. A famous example is the so-called musical dice game, played by Mozart and other eighteenth-century composers, in which a composer cut up a piece of music into small segments (for example, individual measures) and then rolled dice to choose where the segments were placed in the new piece.

EMI, it could be said, was a musical dice game on steroids. To get EMI to create pieces in the style of, say, Mozart, Cope first selected from Mozart's works a large collection of musical segments and applied a computer program he had written that identified key musical patterns he called "signatures"—patterns that help define the composer's unique style. Cope wrote another program that classified each signature as to the particular musical roles it could play in a piece. These signatures were stored in a database corresponding to the composer (Mozart, in our example). Cope also developed in EMI a set of rules—a kind of musical "grammar"—that captured constraints for how variations of signatures could be recombined to create a coherent piece of music in a particular style. EMI employed a random-number generator (the computer equivalent of throwing dice) to select signatures and create musical segments from them; the program then used its musical grammar to help decide how to order the segments.

In this way, EMI could generate a limitless number of new compositions "in the style" of Mozart or any other composer for whom a database of musical signatures had been constructed. Cope carefully chose the best of EMI's compositions to release publicly. I've listened to several of them; to my ear, they range from mediocre to amazingly good, with some beautiful passages, though none have the depth of the original composer's work. (Of course, I say this knowing ahead of time that the pieces are by EMI, so I might be prejudiced.) The longer pieces often contain lovely passages, but also have a non-humanlike tendency to lose the thread of a musical idea. But overall, the published works of EMI were very successful in capturing the style of several different classical composers.

Was EMI creative? My own answer is no. Some of the music EMI generated was quite good, but it was reliant on Cope's musicological knowledge, which was embedded in the musical signatures that Cope curated and the musicological rules he devised. Most crucially, I would contend that the program had no real understanding of the music it generated—neither in terms of the musical concepts nor in terms of the music's emotional impact. For these reasons, EMI could not judge the quality of its own music. That was Cope's job; he said simply, "The works I like are released and the ones that I don't are not."[9]

In 2005, in a decision that I find bewildering, Cope destroyed EMI's entire database of musical signatures. The main reason he gave was that EMI's compositions, being so easily and infinitely producible, were devalued by critics. Cope felt that EMI would be valued as a composer only if it had, as the philosopher Margaret Boden wrote, a "finite oeuvre—as all human composers, beset by mortality, do."[10]

I don't know if my opinion will offer any consolation to Douglas Hofstadter, who was so upset by EMI's most impressive compositions and their ability to fool professional musicians. I understand Hofstadter's worry. As the literary scholar Jonathan Gottschall has observed, "Art is arguably what most distinguishes humans from the rest of creation. It's the thing that makes us proudest of ourselves."[11] But I would add that what makes us proud is not only the generation of art but also our ability to appreciate it, to understand what makes it moving, and to comprehend what it communicates. This appreciation and understanding are essential for both the audience and the artist; without this, I can't call a creation "creative." In short, to answer the question "Could a computer be creative?" I would say yes in principle, but it won't happen soon.

Question: How far are we from creating general human-level AI?

I'll answer this by quoting Oren Etzioni, director of the Allen Institute for AI: "Take your estimate, double it, triple it, quadruple it. That's when."[12]

For a second opinion, recall Andrej Karpathy's assessment from the previous chapter: "We are really, really far away."[13]

That's my view too.

Computers started off as human. In fact, they were usually women who performed calculations by hand or with mechanical desk calculators, such as the calculations needed during World War II to compute missile trajectories to help soldiers aim their artillery guns. This was the original meaning of *computer*. According to Claire Evans's book *Broad Band*, in the 1930s and '40s, "the term 'girl' was used interchangeably with 'computer.' One member of the . . . National Defense Research Committee . . . ball-parked a unit of 'kilogirl' energy as being equivalent to roughly a thousand hours of computing labor."[14]

In the mid-1940s, electronic computers replaced the human kind and immediately became superhuman: unlike any human computer, the machines could calculate "the trajectory of a speeding shell faster than the shell could fly."[15] This was the first of many narrow tasks at which computers have excelled. Today's computers—programmed with state-of-the-art AI algorithms—have conquered many other narrow tasks, but general intelligence still eludes them.

We've seen that over the history of the field well-known AI practitioners have predicted that general AI will arrive in ten years, or fifteen, or twenty-five, or "in a generation." However, none of these predictions has come to pass. As I described in chapter 3, the "long bet" between Ray Kurzweil and Mitchell Kapor, as to whether a program will pass a carefully structured Turing test, will be decided in 2029. My bet is on Kapor;

I wholeheartedly agree with his sentiments, quoted in the prologue: "Human intelligence is a marvelous, subtle, and poorly understood phenomenon. There is no danger of duplicating it anytime soon."[16]

"Prediction is hard, especially about the future." It's debatable who coined this witty saying, but it is as true in AI as in any other domain. Several surveys given to AI practitioners, asking when general AI or "superintelligent" AI will arrive, have exposed a wide spectrum of opinion, ranging from "in the next ten years" to "never."[17] In other words, we don't have a clue.

What we do know is that general human-level AI will require abilities that AI researchers have been struggling for decades to understand and reproduce—commonsense knowledge, abstraction, and analogy, among others—but these abilities have proven to be profoundly elusive. Other major questions remain: Will general AI require consciousness? Having a sense of self? Feeling emotions? Possessing a survival instinct and fear of death? Having a body? As I quoted Marvin Minsky earlier, "This is still a formative period for our ideas about mind."

I find the question of when computers will achieve *superintelligence*— "an intellect that is much smarter than the best human brains in practically every field, including scientific creativity, general wisdom and social skills"[18]—vexing, to say the least.

Several writers have asserted that if computers reach general human-level AI, these machines will quickly become "superintelligent," in a process akin to I. J. Good's vision of an "intelligence explosion" (described in chapter 3). The thinking goes that a computer with general intelligence will be able to read, at lightning speed, all of humanity's documents and learn everything there is to know. Likewise, it will be able to discover, through its ever-increasing deduction abilities, all kinds of new knowledge that it can turn into new cognitive power for itself. Such a machine would not be constrained by the annoying limitations of humans, such as our slowness of thought and learning, our irrationality and cognitive biases, our susceptibility to boredom, our need for sleep, and our emotions, all of which get in the way of productive thinking. In this view, a superintelligent machine would encompass something close to "pure" intelligence, without being constrained by any of our human foibles.

What seems more likely to me is that these supposed limitations of humans are part and parcel of our general intelligence. The cognitive limitations forced upon us by having bodies that work in the world, along with the emotions and "irrational" biases that evolved to allow us to function as a social group, and all the other qualities sometimes considered cognitive "shortcomings," are in fact precisely what enable us to be generally intelligent rather than narrow savants. I can't prove it, but I think it's likely that general intelligence can't be separated from all these apparent shortcomings, in humans or in machines.

In his "Ten Questions and Speculations" section in *GEB* Douglas Hofstadter addressed this issue with a deceptively simple question: "Will a thinking computer be able to add fast?" His answer surprised me when I first read it but now strikes me as correct. "Perhaps not. We ourselves are composed of hardware which does fancy calculations but that doesn't mean that our symbol level, where 'we' are, knows how to carry out the same fancy calculation. Luckily for you, your symbol level (i.e., *you*) can't gain access to the neurons which are doing your thinking—otherwise you'd get addle-brained. . . . Why should it not be the same for an intelligent program?" Hofstadter went on to explain that an intelligent program would, like us, represent numbers as "full-fledged concept[s] the way we do, replete with associations. . . . With all this 'extra baggage' to carry around, an intelligent program will become quite slothful in its adding."[19]

Question: How terrified should we be about AI?

If you rely on movies and science fiction (and even some popular nonfiction) for your view of AI, you'll be afraid of AI becoming conscious, turning malevolent, and trying to enslave or kill us all. But given how far the field seems from achieving anything like general intelligence, this isn't what most people in the AI community worry about. As I've described throughout this book, there are many reasons to worry about our society's headlong dash toward embracing AI technology: the possibility of massive job losses, the potential for misuse of AI systems, and these systems' unreliability and vulnerability to attack—these are just some of the very legitimate worries of people concerned about the impacts of technology on the lives of humans.

I began this book with an account of Douglas Hofstadter's dismay regarding recent AI progress, but he was terrified, for the most part, about something altogether different. Hofstadter's worry was that human cognition and creativity would be too easily matched by AI programs and that the sublime creations of the human minds that he most revered—Chopin, for example—might be rivaled by superficial algorithms like EMI using a "bag of tricks." Hofstadter lamented, "If such minds of infinite subtlety and complexity and emotional depth could be trivialized by a small chip, it would destroy my sense of what humanity is about." Hofstadter was likewise disturbed by Kurzweil's predictions of the oncoming Singularity, agonizing that if Kurzweil was in any way correct, "we will be superseded. We will be relics. We will be left in the dust."

I empathize with Hofstadter on these worries, but I think they are decidedly premature. Above all, the take-home message from this book is that we humans tend to overestimate AI advances and underestimate the complexity of our own intelligence. Today's AI is far from general intelligence, and I don't believe that machine "superintelligence" is anywhere on the horizon. If general AI ever comes about, I am betting that its complexity will rival that of our own brains.

In any ranking of near-term worries about AI, superintelligence should be far down the list. In fact, the opposite of superintelligence is the real problem. Throughout this book, I've described how even the most accomplished AI systems are brittle; that is, they make errors when their input varies too much from the examples on which they've been trained. It's often hard to predict in what circumstances an AI system's brittleness will come to light. In transcribing speech, translating between languages, describing the content of photos, driving in a crowded city—if robust performance is critical, then humans are still needed in the loop. I think the most worrisome aspect of AI systems in the short term is that we will give them too much autonomy without being fully aware of their limitations and vulnerabilities. We tend to anthropomorphize AI systems: we impute human qualities to them and end up overestimating the extent to which these systems can actually be fully trusted.

The economist Sendhil Mullainathan, in writing about the dangers of AI, cited the long-tail phenomenon (which I described in chapter 6) in his notion of "tail risk":

We should be afraid. Not of intelligent machines. But of machines making decisions that they do not have the intelligence to make. I am far more afraid of machine stupidity than of machine intelligence. Machine stupidity creates a tail risk. Machines can make many many good decisions and then one day fail spectacularly on a tail event that did not appear in their training data. This is the difference between specific and general intelligence.[20]

Or as the AI researcher Pedro Domingos so memorably put it, "People worry that computers will get too smart and take over the world, but the real problem is that they're too stupid and they've already taken over the world."[21]

I worry about AI's lack of reliability. I also worry about how it will be used. In addition to the ethical considerations I covered in chapter 7, one particular development that frightens me is the use of AI systems to generate fake media: text, sounds, images, and videos that depict with terrifying realism events that never actually happened.

So, should we be terrified about AI? Yes and no. Superintelligent, conscious machines are not on the horizon. The aspects of our humanity that we most cherish are not going to be matched by "a bag of tricks." At least I don't think so. However, there is a lot to worry about regarding the potential for dangerous and unethical uses of algorithms and data. It's scary, but on the other hand I'm heartened by the wide attention this topic has recently received in the AI community and beyond. There's a sense of cooperation and common purpose that is emerging among researchers, corporations, and politicians on the urgency of reckoning with these issues.

Question: What exciting problems in AI are still unsolved?

Nearly all of them.

When I started working in AI, part of what I found exciting was that nearly all the important questions of the field were open, waiting for new ideas. I think that this is still true.

If we go back to the beginning of the field, the 1955 proposal by John McCarthy and others (described in chapter 1) listed some of AI's major research topics: natural-language processing, neural networks, machine learning, abstract concepts and reasoning, and creativity. In 2015, Microsoft's research director Eric Horvitz joked that "One might even say that the [1955] proposal, if properly reformatted, could be resubmitted to the National Science Foundation . . . today and would probably get some funding by some excited program managers."[22]

This is by no means a criticism of past AI research. Artificial intelligence is at least as hard as any of humanity's other grand scientific challenges. MIT's Rodney Brooks stated this better than anyone else: "When AI got started, the clear inspiration was human level performance and human level intelligence. I think that goal has been what attracted most researchers into the field for the first sixty years. The fact that we do not have anything close to succeeding at those aspirations says not that researchers have not worked hard or have not been brilliant. It says that it is a very hard goal."[23]

The most exciting questions in AI are not only focused on potential applications. The founders of the field were motivated as much by scientific questions about the nature of intelligence as by the desire to develop new technologies. Indeed, the idea that intelligence is a natural phenomenon, one that could be studied like many other phenomena by building simplified computer models, was the motivation that drew many people (including myself) into the field.

The impacts of AI will continue to grow for all of us. I hope that this book has helped you, as a thinking human, to get a sense of the current state of this burgeoning discipline, including its many unsolved problems, the potential risks and benefits of its technologies, and the scientific and philosophical questions it raises for understanding our own human intelligence. And if any computers are reading this, tell me what *it* refers to in the previous sentence and you're welcome to join in the discussion.

Notes

Prologue: Terrified

1. A. Cuthbertson, "DeepMind AlphaGo: AI Teaches Itself 'Thousands of Years of Human Knowledge' Without Help," *Newsweek*, Oct. 18, 2017, www.newsweek.com /deepmind-alphago-ai-teaches-human-help-687620.

2. In the following sections, quotations from Douglas Hofstadter are from a follow-up interview I did with him after the Google meeting; the quotations accurately capture the content and tone of his remarks to the Google group.

3. Jack Schwartz, quoted in G.-C. Rota, *Indiscrete Thoughts* (Boston: Berkhäuser, 1997), 22.

4. D. R. Hofstadter, *Gödel, Escher, Bach: an Eternal Golden Braid* (New York: Basic Books, 1979), 678.

5. Ibid., 676.

6. Quoted in D. R. Hofstadter, "Staring Emmy Straight in the Eye—and Doing My Best Not to Flinch," in *Creativity, Cognition, and Knowledge*, ed. T. Dartnell (Westport, Conn.: Praeger, 2002), 67–100.

7. Quoted in R. Cellan-Jones, "Stephen Hawking Warns Artificial Intelligence Could End Mankind," BBC News, Dec. 2, 2014, www.bbc.com/news/technology-30290540.

8. M. McFarland, "Elon Musk: 'With Artificial Intelligence, We Are Summoning the Demon,'" *Washington Post*, Oct. 24, 2014.

9. Bill Gates, on Reddit, Jan. 28, 2015, www.reddit.com/r/IAmA/comments/2tzjp7/hi _reddit_im_bill_gates_and_im_back_for_my_third/?.

10. Quoted in K. Anderson, "Enthusiasts and Skeptics Debate Artificial Intelligence," *Vanity Fair*, Nov. 26, 2014.

11. R. A. Brooks, "Mistaking Performance for Competence," in *What to Think About Machines That Think*, ed. J. Brockman (New York: Harper Perennial, 2015), 108–11.

12. Quoted in G. Press, "12 Observations About Artificial Intelligence from the O'Reilly AI Conference," *Forbes*, Oct. 31, 2016, www.forbes.com/sites/gilpress/2016/10/31 /12-observations-about-artificial-intelligence-from-the-oreilly-ai-conference/#886a 6012ea2e.

1: The Roots of Artificial Intelligence

1. J. McCarthy et al., "A Proposal for the Dartmouth Summer Research Project in Artificial Intelligence," submitted to the Rockefeller Foundation, 1955, reprinted in *AI Magazine* 27, no. 4 (2006): 12–14.

2. Cybernetics was an interdisciplinary field that studied "control and communication in the animal and in machines." See N. Wiener, *Cybernetics* (Cambridge, Mass.: MIT Press, 1961).

3. Quoted in N. J. Nilsson, *John McCarthy: A Biographical Memoir* (Washington, D.C.: National Academy of Sciences, 2012).

4. McCarthy et al., "Proposal for the Dartmouth Summer Research Project in Artificial Intelligence."

5. Ibid.

6. G. Solomonoff, "Ray Solomonoff and the Dartmouth Summer Research Project in Artificial Intelligence, 1956," accessed Dec. 4, 2018, www.raysolomonoff.com /dartmouth/dartray.pdf.

7. H. Moravic, *Mind Children: The Future of Robot and Human Intelligence* (Cambridge, Mass.: Harvard University Press, 1988), 20.

8. H. A. Simon, *The Shape of Automation for Men and Management* (New York: Harper & Row, 1965), 96. Note that Simon's use of *man* rather than *person* was par for the course in 1960s America.

9. M. L. Minsky, *Computation: Finite and Infinite Machines* (Upper Saddle River, N.J.: Prentice-Hall, 1967), 2.

10. B. R. Redman, *The Portable Voltaire* (New York: Penguin Books, 1977), 225.

11. M. L. Minsky, *The Emotion Machine: Commonsense Thinking, Artificial Intelligence, and the Future of the Human Mind* (New York: Simon & Schuster, 2006), 95.

12. *One Hundred Year Study on Artificial Intelligence (AI100)*, 2016 Report, 13, ai100.stanford .edu/2016-report.

13. Ibid., 12.

14. J. Lehman, J. Clune, and S. Risi, "An Anarchy of Methods: Current Trends in How Intelligence Is Abstracted in AI," *IEEE Intelligent Systems* 29, no. 6 (2014): 56–62.

15. A. Newell and H. A. Simon, "GPS: A Program That Simulates Human Thought," P-2257, Rand Corporation, Santa Monica, Calif. (1961).

16. F. Rosenblatt, "The Perceptron: A Probabilistic Model for Information Storage and Organization in the Brain," *Psychological Review* 65, no. 6 (1958): 386–408.

17. Mathematically, the perceptron-learning algorithm is the following. For each weight w_j: $w_j \leftarrow w_j + \eta\,(t - y)\,x_j$, where t is the correct output (1 or 0) for the given input, y is the actual output of the perceptron, x_j is the input associated with weight w_j, and η is the *learning rate*, a value given by the programmer. The arrow signifies an update. The threshold is incorporated by creating an additional "input" x_0 with a constant value of 1, whose associated weight $w_0 = -threshold$. With this added input and weight (called the bias), the perceptron fires only if the sum of the inputs times weights (that is, the dot product between the input vector and the weight vector) is greater than or equal to 0. Often the input values are scaled and other transformations are applied in order to keep the weights from growing too large.

18. Quoted in M. Olazaran, "A Sociological Study of the Official History of the Perceptrons Controversy," *Social Studies of Science* 26, no. 3 (1996): 611–59.

19. M. A. Boden, *Mind as Machine: A History of Cognitive Science* (Oxford: Oxford University Press, 2006), 2:913.

20. M. L. Minsky and S. L. Papert, *Perceptrons: An Introduction to Computational Geometry* (Cambridge, Mass.: MIT Press, 1969).

21. In technical terms, any Boolean function can be computed by a fully connected multilayer network with linear threshold units and one internal ("hidden") layer.

22. Olazaran, "Sociological Study of the Official History of the Perceptrons Controversy."

23. G. Nagy, "Neural Networks—Then and Now," *IEEE Transactions on Neural Networks* 2, no. 2 (1991): 316–18.

24. Minsky and Papert, *Perceptrons*, 231–32.

25. J. Lighthill, "Artificial Intelligence: A General Survey," in *Artificial Intelligence: A Paper Symposium* (London: Science Research Council, 1973).

26. Quoted in C. Moewes and A. Nürnberger, *Computational Intelligence in Intelligent Data Analysis* (New York: Springer, 2013), 135.

27. M. L. Minsky, *The Society of Mind* (New York: Simon & Schuster, 1987), 29.

2: Neural Networks and the Ascent of Machine Learning

1. The activation value y at each hidden and output unit is typically computed by taking the dot product between the vector x of inputs to the unit and the vector w of weights on the connections to that unit, and applying the sigmoid function to the result: $y = 1/(1 + e^{-(x \cdot w)})$. The x and w vectors also include the "bias" weight and activation. If the units have nonlinear output functions such as sigmoids, with enough hidden units the network can compute any function (with minimal restrictions) to any desired level of approximation. This fact is called the universal approximation theorem. See M. Nielsen, *Neural Networks and Deep Learning*, neuralnetworksanddeeplearning.com, for more details.

2. For readers with some calculus background: Back-propagation is a form of *gradient descent*, which approximates, for each weight w in the network, the direction of steepest descent in the "error surface." This direction is calculated by taking the gradient of the error function (for example, the square of the difference between output and target) with respect to weight w. Consider, for example, the weight w on the connection from input unit i to hidden unit h. Weight w is modified in the direction of steepest descent by an amount determined by the error that has been propagated to unit h as well as the activation of unit i and a user-defined *learning rate*.

For an in-depth explanation of back-propagation, I recommend the free online book by Michael Nielsen, *Neural Networks and Deep Learning*.

3. In my network with 324 inputs, 50 hidden units, and 10 output units, there are $324 \times 50 = 16{,}200$ weights from the inputs to the hidden layer, and $50 \times 10 = 500$ weights from the hidden layer to the output layer, for a total of 16,700 weights.

4. D. E. Rumelhart, J. L. McClelland, and the PDP Research Group, *Parallel Distributed Processing: Explorations in the Microstructure of Cognition* (Cambridge, Mass.: MIT Press, 1986), 1:3.

5. Ibid., 113.

6. Quoted in C. Johnson, "Neural Network Startups Proliferate Across the U.S.," *The Scientist*, Oct. 17, 1988.

7. A. Clark, *Being There: Putting Brain, Body, and World Together Again* (Cambridge, Mass.: MIT Press, 1996), 26.

8. As Douglas Hofstadter pointed out to me, the grammatically correct version is "good old old-fashioned AI," but GOOFAI doesn't have the same ring as GOFAI.

3: AI Spring

1. Q. V. Le et al., "Building High-Level Features Using Large-Scale Unsupervised Learning," in *Proceedings of the International Conference on Machine Learning* (2012), 507–14.

2. P. Hoffman, "Retooling Machine and Man for Next Big Chess Faceoff," *New York Times*, Jan. 21, 2003.

3. D. L. McClain, "Chess Player Says Opponent Behaved Suspiciously," *New York Times*, Sept. 28, 2006.

4. Quoted in M. Y. Vardi, "Artificial Intelligence: Past and Future," *Communications of the Association for Computing Machinery* 55, no. 1 (2012): 5.

5. K. Kelly, "The Three Breakthroughs That Have Finally Unleashed AI on the World," *Wired*, Oct. 27, 2014.

6. J. Despres, "Scenario: Shane Legg," *Future*, accessed Dec. 4, 2018, future.wikia.com /wiki/Scenario:_Shane_Legg.

7. Quoted in H. McCracken, "Inside Mark Zuckerberg's Bold Plan for the Future of Facebook," *Fast Company*, Nov. 16, 2015, www.fastcompany.com/3052885/mark -zuckerberg-facebook.

8. V. C. Müller and N. Bostrom, "Future Progress in Artificial Intelligence: A Survey of Expert Opinion," in *Fundamental Issues of Artificial Intelligence*, ed. V. C. Müller (Cham, Switzerland: Springer International, 2016), 555–72.

9. M. Loukides and B. Lorica, "What Is Artificial Intelligence?," *O'Reilly*, June 20, 2016, www.oreilly.com/ideas/what-is-artificial-intelligence.

10. S. Pinker, "Thinking Does Not Imply Subjugating," in *What to Think About Machines That Think*, ed. J. Brockman (New York: Harper Perennial, 2015), 5–8.

11. A. M. Turing, "Computing Machinery and Intelligence," *Mind* 59, no. 236 (1950): 433–60.

12. J. R. Searle, "Minds, Brains, and Programs," *Behavioral and Brain Sciences* 3, no. 3 (1980): 417–24.

13. J. R. Searle, *Mind: A Brief Introduction* (Oxford: Oxford University Press, 2004), 66.

14. The terms *strong AI* and *weak AI* have also been used to mean something more like *general AI* and *narrow AI*. This is how Ray Kurzweil uses them, but this differs from Searle's original meaning.

15. Searle's article is reprinted in D. R. Hofstadter and D. C. Dennett, *The Mind's I: Fantasies and Reflections on Self and Soul* (New York: Basic Books, 1981), along with a cogent counterargument from Hofstadter.

16. S. Aaronson, *Quantum Computing Since Democritus* (Cambridge, U.K.: Cambridge University Press, 2013), 33.

17. "Turing Test Transcripts Reveal How Chatbot 'Eugene' Duped the Judges," Coventry University, June 30, 2015, www.coventry.ac.uk/primary-news/turing-test-transcripts-reveal-how-chatbot-eugene-duped-the-judges/.

18. "Turing Test Success Marks Milestone in Computing History," University of Reading, June 8, 2014, www.reading.ac.uk/news-and-events/releases/PR583836.aspx.

19. R. Kurzweil, *The Singularity Is Near: When Humans Transcend Biology* (New York: Viking Press, 2005), 7.

20. Ibid., 22–23.

21. I. J. Good, "Speculations Concerning the First Ultraintelligent Machine," *Advances in Computers* 6 (1966): 31–88.

22. V. Vinge, "First Word," *Omni*, Jan. 1983.

23. Kurzweil, *Singularity Is Near*, 241, 317, 198–99.

24. B. Wang, "Ray Kurzweil Responds to the Issue of Accuracy of His Predictions," *Next Big Future*, Jan. 19, 2010, www.nextbigfuture.com/2010/01/ray-kurzweil-responds-to-issue-of.html.

25. D. Hochman, "Reinvent Yourself: The Playboy Interview with Ray Kurzweil," *Playboy*, April 19, 2016, www.playboy.com/articles/playboy-interview-ray-kurzweil.

26. Kurzweil, *Singularity Is Near*, 136.

27. A. Kreye, "A John Henry Moment," in Brockman, *What to Think About Machines That Think*, 394–96.

28. Kurzweil, *Singularity Is Near*, 494.

29. R. Kurzweil, "A Wager on the Turing Test: Why I Think I Will Win," Kurzweil AI, April 9, 2002, www.kurzweilai.net/a-wager-on-the-turing-test-why-i-think-i-will-win.

30. Ibid.

31. Ibid.

32. Ibid.

33. M. Dowd, "Elon Musk's Billion-Dollar Crusade to Stop the A.I. Apocalypse," *Vanity Fair*, March 26, 2017.

34. L. Grossman, "2045: The Year Man Becomes Immortal," *Time*, Feb. 10, 2011.

35. From Singularity University website, accessed Dec. 4, 2018, su.org/about/.

36. Kurzweil, *Singularity Is Near*, 316.

37. R. Kurzweil, *The Age of Spiritual Machines: When Computers Exceed Human Intelligence* (New York: Viking Press, 1999), 170.

38. D. R. Hofstadter, "Moore's Law, Artificial Evolution, and the Fate of Humanity," in *Perspectives on Adaptation in Natural and Artificial Systems*, ed. L. Booker et al. (New York: Oxford University Press, 2005), 181.

39. All of these quotations are from Kurzweil, *Age of Spiritual Machines*, 169–70.

40. Hofstadter, "Moore's Law, Artificial Evolution, and the Fate of Humanity," 182.

41. From Long Bets website: longbets.org/about.

42. From Long Bets website, Bet 1: longbets.org/1/#adjudication_terms.

43. Ibid.

44. Ibid.

45. Kurzweil, "Wager on the Turing Test."
46. M. Kapor, "Why I Think I Will Win," Kurzweil AI, April 9, 2002, http://www
    .kurzweilai.net/why-i-think-i-will-win.
47. Ibid.
48. R. Kurzweil, foreword to *Virtual Humans*, by P. M. Plantec (New York: AMACOM, 2004).
49. Quoted in Grossman, "2045."

4: Who, What, When, Where, Why

1. S. A. Papert, "The Summer Vision Project," MIT Artificial Intelligence Group Vision
   Memo 100 (July 7, 1966), dspace.mit.edu/handle/1721.1/6125.
2. D. Crevier, *AI: The Tumultuous History of the Search for Artificial Intelligence* (New York:
   Basic Books, 1993), 88.
3. K. Fukushima, "Cognitron: A Self-Organizing Multilayered Neural Network Model,"
   Biological Cybernetics 20, no. 3–4 (1975): 121–36; K. Fukushima, "Neocognitron: A Hierarchical Neural Network Capable of Visual Pattern Recognition," *Neural Networks*
   1, no. 2 (1988): 119–30.
4. Before being fed to the network, the image needs to be scaled to a fixed size—the
   same size as the network's first layer.
5. Most claims about how the brain performs some task have to come with many caveats;
   the story I've just outlined is no different. While what I've said is approximately accurate, the brain is outrageously complex, and the findings I've outlined are only a small
   part of the story of early vision, much of which scientists still don't fully understand.
6. The array of weights associated with each activation map is called a convolutional
   filter or convolutional kernel.
7. Here I'm using the term *classification module* as shorthand for what are usually called
   the fully connected layers of a deep convolutional network.
8. My description of ConvNets leaves out many details. For example, to compute its
   activation, a unit in a convolutional layer performs a convolution and then applies
   a nonlinear activation function to the result. ConvNets also typically feature other
   types of layers, such as "pooling layers." For details, see I. Goodfellow, Y. Bengio,
   and A. Courville, *Deep Learning* (Cambridge, Mass.: MIT Press, 2016).
9. At the time of this writing, Google's "search by image" engine is accessed at images
   .google.com by clicking the small camera icon in the search box.

5: ConvNets and ImageNet

1. Indeed, back-propagation is an algorithm that was discovered independently by
   several different groups, and—ironically, given back-propagation's function as a
   credit-assignment algorithm—assigning the credit for its discovery has been a long-
   standing battle among neural network researchers.
2. Quoted in D. Hernandez, "Facebook's Quest to Build an Artificial Brain Depends on
   This Guy," *Wired*, Aug. 14, 2014, www.wired.com/2014/08/deep-learning-yann-lecun/.
3. There was also a "detection" competition, in which programs had to also locate objects of the various categories in images, as well as other specialized challenges; here
   I'm focusing on the classification challenge.
4. D. Gershgorn, "The Data That Transformed AI Research—and Possibly the World,"
   Quartz, July 26, 2017, qz.com/1034972/the-data-that-changed-the-direction-of-ai
   -research-and-possibly-the-world/.

5. "About Amazon Mechanical Turk," www.mturk.com/help.

6. L. Fei-Fei and J. Deng, "ImageNet: Where Have We Been? Where Are We Going?," slides at image-net.org/challenges/talks_2017/imagenet_ilsvrc2017_v1.0.pdf.

7. A. Krizhevsky, I. Sutskever, and G. E. Hinton, "ImageNet Classification with Deep Convolutional Neural Networks," *Advances in Neural Information Processing Systems 25* (2012): 1097–105.

8. T. Simonite, "Teaching Machines to Understand Us," *Technology Review*, Aug. 5, 2015, www.technologyreview.com/s/540001/teaching-machines-to-understand-us/.

9. ImageNet Large Scale Visual Recognition Challenge announcement, June 2, 2015, www.image-net.org/challenges/LSVRC/announcement-June-2-2015.

10. S. Chen, "Baidu Fires Scientist Responsible for Breaching Rules in High-Profile Super-computer AI Test," *South China Morning Post*, international edition, June 12, 2015, www .scmp.com/tech/science-research/article/1820649/chinas-baidu-fires-researcher-after -team-cheated-high-profile.

11. Gershgorn, "Data That Transformed AI Research."

12. Quoted in Hernandez, "Facebook's Quest to Build an Artificial Brain Depends on This Guy."

13. B. Agüera y Arcas, "Inside the Machine Mind: Latest Insights on Neuroscience and Computer Science from Google" (lecture video), Oxford Martin School, May 10, 2016, www.youtube.com/watch?v=v1dW7ViahEc.

14. K. He et al., "Delving Deep into Rectifiers: Surpassing Human-Level Performance on ImageNet Classification," in *Proceedings of the IEEE International Conference on Computer Vision* (2015), 1026–34.

15. A. Linn, "Microsoft Researchers Win ImageNet Computer Vision Challenge," *AI Blog*, Microsoft, Dec. 10, 2015, blogs.microsoft.com/ai/2015/12/10/microsoft-researchers -win-imagenet-computer-vision-challenge.

16. A. Hern, "Computers Now Better than Humans at Recognising and Sorting Images," *Guardian*, May 13, 2015, www.theguardian.com/global/2015/may/13/baidu -minwa-supercomputer-better-than-humans-recognising-images; T. Benson, "Microsoft Has Developed a Computer System That Can Identify Objects Better than Humans," UPI, Feb. 14, 2015, www.upi.com/Science_News/2015/02/14/Microsoft-has-developed -a-computer-system-that-can-identify-objects-better-than-humans/1171423959603.

17. A. Karpathy, "What I Learned from Competing Against a ConvNet on ImageNet," Sept. 2, 2014, karpathy.github.io/2014/09/02/what-i-learned-from-competing-against -a-convnet-on-imagenet.

18. S. Lohr, "A Lesson of Tesla Crashes? Computer Vision Can't Do It All Yet," *New York Times*, Sept. 19, 2016.

6: A Closer Look at Machines That Learn

1. Readers who followed the 2016 U.S. presidential election will recognize the pun on Bernie Sanders's supporters' tagline, "Feel the Bern."

2. E. Brynjolfsson and A. McAfee, "The Business of Artificial Intelligence," *Harvard Business Review*, July 2017.

3. O. Tanz, "Can Artificial Intelligence Identify Pictures Better than Humans?," *Entrepreneur*, April 1, 2017, www.entrepreneur.com/article/283990.

4. D. Vena, "3 Top AI Stocks to Buy Now," *Motley Fool*, March 27, 2017, www.fool.com /investing/2017/03/27/3-top-ai-stocks-to-buy-now.aspx.

5. Quoted in C. Metz, "A New Way for Machines to See, Taking Shape in Toronto," *New York Times*, Nov. 28, 2017, www.nytimes.com/2017/11/28/technology/artificial -intelligence-research-toronto.html.

6. Quoted in J. Tanz, "Soon We Won't Program Computers. We'll Train Them Like Dogs," *Wired*, May 17, 2016.

7. From Harry Shum's lecture at the Microsoft Faculty Summit, Redmond, Wash., June 2017.

8. An in-depth discussion of this issue is given in J. Lanier, *Who Owns the Future?* (New York: Simon & Schuster, 2013).

9. Tesla's *Customer Privacy Policy*, accessed Dec. 7, 2018, www.tesla.com/about/legal.

10. T. Bradshaw, "Self-Driving Cars Prove to Be Labour-Intensive for Humans," *Financial Times*, July 8, 2017.

11. "Ground Truth Datasets for Autonomous Vehicles," Mighty AI, accessed Dec. 7, 2018, mty.ai/adas/.

12. "Deep Learning in Practice: Speech Recognition and Beyond," EmTech Digital video, May 23, 2016, events.technologyreview.com/emtech/digital/16/video/watch /andrew-ng-deep-learning.

13. Y. Bengio, "Machines That Dream," in *The Future of Machine Intelligence: Perspectives from Leading Practitioners*, ed. D. Beyer (Sebastopol, Calif.: O'Reilly Media), 14.

14. W. Landecker et al., "Interpreting Individual Classifications of Hierarchical Networks," in *Proceedings of the 2013 IEEE Symposium on Computational Intelligence and Data Mining* (2013), 32–38.

15. M. R. Loghmani et al., "Recognizing Objects in-the-Wild: Where Do We Stand?," in *IEEE International Conference on Robotics and Automation* (2018), 2170–77.

16. H. Hosseini et al., "On the Limitation of Convolutional Neural Networks in Recognizing Negative Images," in *Proceedings of the 16th IEEE International Conference on Machine Learning and Applications* (2017), 352–58; R. Geirhos et al., "Generalisation in Humans and Deep Neural Networks," *Advances in Neural Information Processing Systems* 31 (2018): 7549–61; M. Alcorn et al., "Strike (with) a Pose: Neural Networks Are Easily Fooled by Strange Poses of Familiar Objects," arXiv:1811.11553 (2018).

17. M. Orcutt, "Are Face Recognition Systems Accurate? Depends on Your Race," *Technology Review*, July 6, 2016, www.technologyreview.com/s/601786/are-face-recognition -systems-accurate-depends-on-your-race.

18. J. Zhao et al., "Men Also Like Shopping: Reducing Gender Bias Amplification Using Corpus-Level Constraints," in *Proceedings of the 2017 Conference on Empirical Methods in Natural Language Processing* (2017).

19. W. Knight, "The Dark Secret at the Heart of AI," *Technology Review*, April 11, 2017, www.technologyreview.com/s/604087/the-dark-secret-at-the-heart-of-ai/.

20. C. Szegedy et al., "Intriguing Properties of Neural Networks," in *Proceedings of the International Conference on Learning Representations* (2014).

21. A. Nguyen, J. Yosinski, and J. Clune, "Deep Neural Networks Are Easily Fooled: High Confidence Predictions for Unrecognizable Images," in *Proceedings of the IEEE Conference on Computer Vision and Pattern Recognition* (2015), 427–36.

22. See, for example, M. Mitchell, *An Introduction to Genetic Algorithms* (Cambridge, Mass.: MIT Press, 1996).

23. Nguyen, Yosinski, and Clune, "Deep Neural Networks Are Easily Fooled."

24. M. Sharif et al., "Accessorize to a Crime: Real and Stealthy Attacks on State-of-the-Art Face Recognition," in *Proceedings of the 2016 ACM SIGSAC Conference on Computer and Communications Security* (2016), 1528–40.

25. K. Eykholt et al., "Robust Physical-World Attacks on Deep Learning Visual Classification," in *Proceedings of the IEEE Conference on Computer Vision and Pattern Recognition* (2018), 1625–34.

26. S. G. Finlayson et al., "Adversarial Attacks on Medical Machine Learning," *Science* 363, no. 6433 (2019): 1287–89.

27. Quoted in W. Knight, "How Long Before AI Systems Are Hacked in Creative New Ways?," *Technology Review*, Dec. 15, 2016, www.technologyreview.com/s/603116 /how-long-before-ai-systems-are-hacked-in-creative-new-ways.

28. J. Clune, "How Much Do Deep Neural Networks Understand About the Images They Recognize?," lecture slides (2016), accessed Dec. 7, 2018, c4dm.eecs.qmul.ac.uk /horse2016/HORSE2016_Clune.pdf.

7: On Trustworthy and Ethical AI

1. Quoted in D. Palmer, "AI Could Help Solve Humanity's Biggest Issues by Taking Over from Scientists, Says DeepMind CEO," *Computing*, May 26, 2015, www .computing.co.uk/ctg/news/2410022/ai-could-help-solve-humanity-s-biggest-issues -by-taking-over-from-scientists-says-deepmind-ceo.

2. S. Lynch, "Andrew Ng: Why AI Is the New Electricity," *Insights by Stanford Business*, March 11, 2017, www.gsb.stanford.edu/insights/andrew-ng-why-ai-new-electricity.

3. J. Anderson, L. Rainie, and A. Luchsinger, "Artificial Intelligence and the Future of Humans," Pew Research Center, Dec. 10, 2018, www.pewinternet.org/2018/12/10 /artificial-intelligence-and-the-future-of-humans.

4. Two recent treatments of the ethical issues surrounding AI and big data are C. O'Neil, *Weapons of Math Destruction: How Big Data Increases Inequality and Threatens Democracy* (New York: Crown, 2016), and H. Fry, *Hello World: Being Human in the Age of Algorithms* (New York: W. W. Norton, 2018).

5. C. Domonoske, "Facebook Expands Use of Facial Recognition to ID Users in Photos," National Public Radio, Dec. 19, 2017, www.npr.org/sections/thetwo-way/2017 /12/19/571954455/facebook-expands-use-of-facial-recognition-to-id-users-in-photos.

6. H. Hodson, "Face Recognition Row over Right to Identify You in the Street," *New Scientist*, June 19, 2015.

7. J. Snow, "Amazon's Face Recognition Falsely Matched 28 Members of Congress with Mugshots," *Free Future* (blog), ACLU, July 26, 2018, www.aclu.org/blog /privacy-technology/surveillance-technologies/amazons-face-recognition-falsely -matched-28.

8. B. Brackeen, "Facial Recognition Software Is Not Ready for Use by Law Enforcement," *Tech Crunch*, June 25, 2018, techcrunch.com/2018/06/25/facial-recognition-software -is-not-ready-for-use-by-law-enforcement.

9. B. Smith, "Facial Recognition Technology: The Need for Public Regulation and Corporate Responsibility," *Microsoft on the Issues* (blog), Microsoft, July 13, 2018, blogs.microsoft.com/on-the-issues/2018/07/13/facial-recognition-technology-the -need-for-public-regulation-and-corporate-responsibility.

10. K. Walker, "AI for Social Good in Asia Pacific," *Around the Globe* (blog), Google, Dec. 13, 2018, www.blog.google/around-the-globe/google-asia/ai-social-good-asia-pacific.

11. B. Goodman and S. Flaxman, "European Union Regulations on Algorithmic Decision-Making and a 'Right to Explanation,'" *AI Magazine* 38, no. 3 (Fall 2017): 50–57.

12. "Article 12, EU GDPR: Transparent Information, Communication, and Modalities for the Exercise of the Rights of the Data Subject," EU General Data Protection Regulation, accessed Dec. 7, 2018, www.privacy-regulation.eu/en/article-12-transparent-information-communication-and-modalities-for-the-exercise-of-the-rights-of-the-data-subject-GDPR.htm.

13. Partnership on AI website, accessed Dec. 18, 2018, www.partnershiponai.org.

14. For an extended survey of this topic, see W. Wallach and C. Allen, *Moral Machines: Teaching Robots Right from Wrong* (New York: Oxford University Press, 2008).

15. I. Asimov, *I, Robot* (Bantam Dell, 2004), 37. (First edition: Grove, 1950.)

16. A. C. Clarke, *2001: A Space Odyssey* (London: Hutchinson & Co, 1968).

17. Ibid., 192.

18. N. Wiener, "Some Moral and Technical Consequences of Automation," *Science* 131, no. 3410 (1960): 1355–58.

19. J. J. Thomson, "The Trolley Problem," *Yale Law Journal* 94, no. 6 (1985): 1395–415.

20. For example, see J. Achenbach, "Driverless Cars Are Colliding with the Creepy Trolley Problem," *Washington Post*, December 29, 2015.

21. J.-F. Bonnefon, A. Shariff, and I. Rahwan, "The Social Dilemma of Autonomous Vehicles," *Science* 352, no. 6293 (2016): 1573–76.

22. J. D. Greene, "Our Driverless Dilemma," *Science* 352, no. 6293 (2016): 1514–15.

23. For example, see M. Anderson and S. L. Anderson, "Machine Ethics: Creating an Ethical Intelligent Agent," *AI Magazine* 28, no. 4 (2007): 15.

8: Rewards for Robots

1. A. Sutherland, "What Shamu Taught Me About a Happy Marriage," *New York Times*, June 25, 2006, www.nytimes.com/2006/06/25/fashion/what-shamu-taught-me-about-a-happy-marriage.html.

2. thejetsons.wikia.com/wiki/Rosey.

3. To be more precise, this approach to reinforcement learning, called value learning, is not the only possible approach. A second approach, called policy learning, has the goal of learning directly what action to perform in a given state, rather than first learning the numerical *values* of actions.

4. C. J. Watkins and P. Dayan, "Q-Learning," *Machine Learning* 8, nos. 3–4 (1992): 279–92.

5. For a detailed, technical introduction to reinforcement learning, see R. S. Sutton and A. G. Barto, *Reinforcement Learning: An Introduction*, 2nd ed. (Cambridge, Mass.: MIT Press, 2017), incompleteideas.net/book/the-book-2nd.html.

6. For example, see the following papers: P. Christiano et al., "Transfer from Simulation to Real World Through Learning Deep Inverse Dynamics Model," arXiv:1610.03518 (2016); J. P. Hanna and P. Stone, "Grounded Action Transformation for Robot Learning in Simulation," in *Proceedings of the Conference of the American Association for Artificial Intelligence* (2017), 3834–40; A. A. Rusu et al., "Sim-to-Real Robot Learning from Pixels with Progressive Nets," in *Proceedings of the First Annual Conference on Robot Learning, CoRL* (2017); S. James, A. J. Davison, and E. Johns, "Transferring End-to-End Visuomotor Control from Simulation to Real World for a Multi-stage Task," in *Proceedings of the First Annual Conference on Robot Learning, CoRL* (2017); M. Cutler, T. J. Walsh, and J. P. How, "Real-World Re-

inforcement Learning via Multifidelity Simulators," *IEEE Transactions on Robotics* 31, no. 3 (2015): 655–71.

9: Game On

1. Demis Hassabis, quoted in P. Iwaniuk, "A Conversation with Demis Hassabis, the Bullfrog AI Prodigy Now Finding Solutions to the World's Big Problems," *PCGamesN*, accessed Dec. 7, 2018, www.pcgamesn.com/demis-hassabis-interview.
2. Quoted in "From Not Working to Neural Networking," *Economist*, June 25, 2016.
3. M. G. Bellemare et al., "The Arcade Learning Environment: An Evaluation Platform for General Agents," *Journal of Artificial Intelligence Research* 47 (2013): 253–79.
4. More technically, DeepMind's program used what is called an epsilon-greedy method for choosing an action at each time step. With probability *epsilon* the program chooses an action at random; with probability (1 − *epsilon*) the program chooses the action with the highest value. *Epsilon* is a value between 0 and 1; it is initially set close to 1 and is gradually decreased over the episodes of training.
5. R. S. Sutton and A. G. Barto, *Reinforcement Learning: An Introduction*, 2nd ed. (Cambridge, Mass.: MIT Press, 2017), 124, incompleteideas.net/book/the-book-2nd.html.
6. For more details, see V. Mnih et al., "Human-Level Control Through Deep Reinforcement Learning," *Nature* 518, no. 7540 (2015): 529.
7. V. Mnih et al., "Playing Atari with Deep Reinforcement Learning," *Proceedings of the Neural Information Processing Systems (NIPS) Conference, Deep Learning Workshop* (2013).
8. "Arthur Samuel," History of Computers website, history-computer.com/Modern Computer/thinkers/Samuel.html.
9. Samuel's program used a variable number of plies, depending on the move.
10. Samuel's program also used a method called alpha-beta pruning at each turn to determine nodes in the game tree that did not need to be evaluated. Alpha-beta pruning was also an essential part of IBM's Deep Blue chess-playing program.
11. For details, see A. L. Samuel, "Some Studies in Machine Learning Using the Game of Checkers," *IBM Journal of Research and Development* 3, no. 3 (1959): 210–29.
12. Ibid.
13. J. Schaeffer et al., "CHINOOK: The World Man-Machine Checkers Champion," *AI Magazine* 17, no. 1 (1996): 21.
14. D. Hassabis, "Artificial Intelligence: Chess Match of the Century," *Nature* 544 (2017): 413–14.
15. A. Newell, J. Calman Shaw, and H. A. Simon, "Chess-Playing Programs and the Problem of Complexity," *IBM Journal of Research and Development* 2, no. 4 (1958): 320–35.
16. M. Newborn, *Deep Blue: An Artificial Intelligence Milestone* (New York: Springer, 2003), 236.
17. Quoted in J. Goldsmith, "The Last Human Chess Master," *Wired*, Feb. 1, 1995.
18. Quoted in M. Y. Vardi, "Artificial Intelligence: Past and Future," *Communications of the Association for Computing Machinery* 55, no. 1 (2012): 5.
19. A. Levinovitz, "The Mystery of Go, the Ancient Game That Computers Still Can't Win," *Wired*, May 12, 2014.
20. G. Johnson, "To Test a Powerful Computer, Play an Ancient Game," *New York Times*, July 29, 1997.
21. Quoted in "S. Korean Go Player Confident of Beating Google's AI," Yonhap News Agency, Feb. 23, 2016, english.yonhapnews.co.kr/search1/2603000000.html?cid=AEN 20160223003651315.

22. Quoted in M. Zastrow, "'I'm in Shock!': How an AI Beat the World's Best Human at Go," *New Scientist*, March 9, 2016, www.newscientist.com/article/2079871-im-in -shock-how-an-ai-beat-the-worlds-best-human-at-go.

23. C. Metz, "The Sadness and Beauty of Watching Google's AI Play Go," *Wired*, March 11, 2016, www.wired.com/2016/03/sadness-beauty-watching-googles-ai-play-go.

24. "For Artificial Intelligence to Thrive, It Must Explain Itself," *Economist*, Feb. 15, 2018, www.economist.com/news/science-and-technology/21737018-if-it-cannot-who-will -trust-it-artificial-intelligence-thrive-it-must.

25. P. Taylor, "The Concept of 'Cat Face,'" *London Review of Books*, Aug. 11, 2016.

26. Quoted in S. Byford, "DeepMind Founder Demis Hassabis on How AI Will Shape the Future," *Verge*, March 10, 2016, www.theverge.com/2016/3/10/11192774/demis -hassabis-interview-alphago-google-deepmind-ai.

27. D. Silver et al., "Mastering the Game of Go Without Human Knowledge," *Nature*, 550 (2017): 354–59.

28. D. Silver et al., "A General Reinforcement Learning Algorithm That Masters Chess, Shogi, and Go Through Self-Play," *Science* 362, no. 6419 (2018): 1140–44.

10: Beyond Games

1. Quoted in P. Iwaniuk, "A Conversation with Demis Hassabis, the Bullfrog AI Prodigy Now Finding Solutions to the World's Big Problems," *PCGamesN*, accessed Dec. 7, 2018, www.pcgamesn.com/demis-hassabis-interview.

2. E. David, "DeepMind's AlphaGo Mastered Chess in Its Spare Time," Silicon Angle, Dec. 6, 2017, siliconangle.com/blog/2017/12/06/deepminds-alphago-mastered-chess -spare-time.

3. As one example, still in the game-playing domain, DeepMind published a paper in 2018 describing a reinforcement-learning system that they claimed exhibited some degree of transfer learning in its ability to play different Atari games. L. Espeholt et al., "Impala: Scalable Distributed Deep-RL with Importance Weighted Actor-Learner Architectures," in *Proceedings of the International Conference on Machine Learning* (2018), 1407–16.

4. D. Silver et al., "Mastering the Game of Go Without Human Knowledge," *Nature* 550 (2017): 354–59.

5. G. Marcus, "Innateness, AlphaZero, and Artificial Intelligence," arXiv:1801.05667 (2018).

6. F. P. Such et al., "Deep Neuroevolution: Genetic Algorithms Are a Competitive Alternative for Training Deep Neural Networks for Reinforcement Learning," *Proceedings of the Neural Information Processing Systems (NIPS) Conference, Deep Reinforcement Learning Workshop* (2018).

7. M. Mitchell, *An Introduction to Genetic Algorithms* (Cambridge, Mass.: MIT Press, 1996).

8. Marcus, "Innateness, AlphaZero, and Artificial Intelligence."

9. G. Marcus, "Deep Learning: A Critical Appraisal," arXiv:1801.00631 (2018).

10. K. Kansky et al., "Schema Networks: Zero-Shot Transfer with a Generative Causal Model of Intuitive Physics," in *Proceedings of the International Conference on Machine Learning* (2017), 1809–18.

11. A. A. Rusu et al., "Progressive Neural Networks," arXiv:1606.04671 (2016).

12. Marcus, "Deep Learning."

13. Quoted in N. Sonnad and D. Gershgorn, "Q&A: Douglas Hofstadter on Why AI Is Far from Intelligent," *Quartz*, Oct. 10, 2017, qz.com/1088714/qa-douglas-hofstadter-on-why-ai-is-far-from-intelligent.

14. I should note that a few robotics groups have actually developed dishwasher-loading robots, though none of these was trained by reinforcement learning, or any other kind of machine-learning method, as far as I know. These robots come with some impressive videos (for example, "Robotic Dog Does Dishes, Plays Fetch," NBC New York, June 23, 2016, www.nbcnewyork.com/news/local/Boston-Dynamics-Dog-Does-Dishes-Brings-Sodas-384140021.html), but it's clear that they are still quite limited and not yet ready to solve my family's nightly dishwashing arguments.

15. A. Karpathy, "AlphaGo, in Context," *Medium*, May 31, 2017, medium.com/@karpathy/alphago-in-context-c47718cb95a5.

11: Words, and the Company They Keep

1. My "Restaurant" story was inspired by similar tiny stories created by Roger Schank and his colleagues in their work on natural language understanding (R. C. Schank and C. K. Riesbeck, *Inside Computer Understanding: Five Programs Plus Miniatures* [Hillsdale, N.J.: Lawrence Erlbaum Associates, 1981]) and by John Searle in his critiques of AI (J. R. Searle, "Minds, Brains, and Programs," *Behavioral and Brain Sciences* 3, no. 3 [1980]: 417–24).

2. G. Hinton et al., "Deep Neural Networks for Acoustic Modeling in Speech Recognition: The Shared Views of Four Research Groups," *IEEE Signal Processing Magazine* 29, no. 6 (2012): 82–97.

3. J. Dean, "Large Scale Deep Learning," slides from keynote lecture, Conference on Information and Knowledge Management (CIKM), Nov. 2014, accessed Dec. 7, 2018, static.googleusercontent.com/media/research.google.com/en//people/jeff/CIKM-keynote-Nov2014.pdf.

4. S. Levy, "The iBrain Is Here, and It's Already in Your Phone," *Wired*, Aug. 24, 2016, www.wired.com/2016/08/an-exclusive-look-at-how-ai-and-machine-learning-work-at-apple.

5. In the speech-recognition literature, the most commonly used performance metric is "word-error rate" on large collections of short audio segments. While the word-error-rate performance of state-of-the-art speech-recognition systems applied to these collections is at or above "human level," there are several reasons to argue that when more realistic measures are used (for example, noisy or accented speech, important words, ambiguous language), speech-recognition performance by machines is still significantly below that of humans. A good overview of some of these arguments is given in A. Hannun, "Speech Recognition Is Not Solved," accessed Dec. 7, 2018, awni.github.io/speech-recognition.

6. A good, though technical, overview of how modern speech-recognition algorithms work is given in J.H.L. Hansen and T. Hasan, "Speaker Recognition by Machines and Humans: A Tutorial Review," *IEEE Signal Processing Magazine* 32, no. 6 (2015): 74–99.

7. These reviews are from Amazon.com; in some cases, I have lightly edited them.

8. At the time of this writing, the online world is still reeling from the news that a data-analytics company called Cambridge Analytica used data from tens of millions of

Facebook accounts to help target political ads, likely using sentiment-classification methods, among other techniques.

9. Recall from chapter 2 that each unit in a neural network computes a mathematical function of the sum of its inputs times their weights. This can be done only if the inputs are numbers.

10. J. Firth, "A Synopsis of Linguistic Theory, 1930–1955," in *Studies in Linguistic Analysis* (Oxford: Philological Society, 1957), 1–32.

11. A. Lenci, "Distributional Semantics in Linguistic and Cognitive Research," *Italian Journal of Linguistics* 20, no. 1 (2008): 1–31.

12. In physics, the term *vector* is often defined as an entity having a magnitude and direction. This definition is equivalent to the one I gave in the text: any vector can be uniquely described by the coordinates of a point, where the magnitude is the length of a segment from the origin to that point, and the direction is the angle this segment makes with the coordinate axes.

13. T. Mikolov et al., "Efficient Estimation of Word Representations in Vector Space," in *Proceedings of the International Conference on Learning Representations* (2013).

14. Word2vec, Google Code Archive, code.google.com/archive/p/word2vec/. Word vectors are also called word embeddings.

15. Here, I'm illustrating a version of the "skip-gram" method, which was one of two methods proposed in Mikolov et al., "Efficient Estimation of Word Representations in Vector Space."

16. Ibid.

17. I used the word2vec demo at bionlp-www.utu.fi/wv_demo/ (using "English GoogleNews Negative300" model) to obtain these results.

18. The idea is to solve for x in the vector arithmetic problem $man - woman = king - x$. To add or subtract two vectors, just add or subtract their corresponding elements; for example, $(3, 2, 4) - (1, 1, 1) = (2, 1, 3)$.

19. bionlp-www.utu.fi/wv_demo/.

20. R. Kiros et al., "Skip-Thought Vectors," in *Advances in Neural Information Processing Systems* 28 (2015), 3294–302.

21. Quoted in H. Devlin, "Google a Step Closer to Developing Machines with Human-Like Intelligence," *Guardian*, May 21, 2015, www.theguardian.com/science/2015/may/21/google-a-step-closer-to-developing-machines-with-human-like-intelligence.

22. Y. LeCun, "What's Wrong with Deep Learning?," lecture slides, p. 77, accessed Dec. 14, 2018, www.pamitc.org/cvpr15/files/lecun-20150610-cvpr-keynote.pdf.

23. For example, see T. Bolukbasi et al., "Man Is to Computer Programmer as Woman Is to Homemaker? Debiasing Word Embeddings," in *Advances in Neural Information Processing Systems* 29 (2016), 4349–57.

24. For example, see J. Zhao et al., "Learning Gender-Neutral Word Embeddings," in *Proceedings of the 2018 Conference on Empirical Methods in Natural Language Processing* (2018), 4847–53, and A. Sutton, T. Lansdall-Welfare, and N. Cristianini, "Biased Embeddings from Wild Data: Measuring, Understanding, and Removing," in *Proceedings of the International Symposium on Intelligent Data Analysis* (2018), 328–39.

12: Translation as Encoding and Decoding

1. Q. V. Le and M. Schuster, "A Neural Network for Machine Translation, at Production Scale," *AI Blog*, Google, Sept. 27, 2016, ai.googleblog.com/2016/09/a-neural-network-for-machine.html.

2. W. Weaver, "Translation," in *Machine Translation of Languages*, ed. W. N. Locke and A. D. Booth (New York: Technology Press and John Wiley & Sons, 1955), 15–23.

3. This is the method used by Google Translate for most languages. At the time of this writing, Google Translate has not yet switched to neural networks for some less common languages.

4. For more details, see Y. Wu et al., "Google's Neural Machine Translation System: Bridging the Gap Between Human and Machine Translation," arXiv:1609.08144 (2016).

5. In Google's neural machine-translation system, the word vectors are learned as part of the training of the entire network.

6. More specifically, the outputs of the decoder network are probabilities for each possible word in the network's vocabulary (here, French). More details are given in Wu et al., "Google's Neural Machine Translation System."

7. At the time of this writing, Google Translate and other translation systems work by translating one sentence at a time. An example of research on going beyond sentence-by-sentence translation is described in L. M. Werlen and A. Popescu-Belis, "Using Coreference Links to Improve Spanish-to-English Machine Translation," in *Proceedings of the 2nd Workshop on Coreference Resolution Beyond OntoNotes* (2017), 30–40.

8. S. Hochreiter and J. Schmidhuber, "Long Short-Term Memory," *Neural Computation* 9, no. 8 (1997): 1735–80.

9. Wu et al., "Google's Neural Machine Translation System."

10. Ibid.

11. T. Simonite, "Google's New Service Translates Languages Almost as Well as Humans Can," *Technology Review*, Sept. 27, 2016, www.technologyreview.com/s/602480/googles -new-service-translates-languages-almost-as-well-as-humans-can.

12. A. Linn, "Microsoft Reaches a Historic Milestone, Using AI to Match Human Performance in Translating News from Chinese to English," *AI Blog*, Microsoft, March 14, 2018, blogs.microsoft.com/ai/machine-translation-news-test-set-human-parity.

13. "IBM Watson Is Now Fluent in Nine Languages (and Counting)," *Wired*, Oct. 6, 2016, www.wired.co.uk/article/connecting-the-cognitive-world.

14. A. Packer, "Understanding the Language of Facebook," EmTech Digital video lecture, May 23, 2016, events.technologyreview.com/video/watch/alan-packer-understanding -language.

15. DeepL Pro, press release, March 20, 2018, www.deepl.com/press.html.

16. K. Papineni et al., "BLEU: A Method for Automatic Evaluation of Machine Translation," in *Proceedings of the 40th Annual Meeting of the Association for Computational Linguistics* (2002), 311–18.

17. Wu et al., "Google's Neural Machine Translation System"; H. Hassan et al., "Achieving Human Parity on Automatic Chinese to English News Translation," arXiv:1803.05567 (2018).

18. Google Translate's French translation of the "Restaurant" story: Un homme est entré dans un restaurant et a commandé un hamburger, cuit rare. Quand il est arrivé, il a été brûlé à un croustillant. La serveuse s'arrêta devant la table de l'homme. "Est-ce que le hamburger va bien?" Demanda-t-elle. "Oh, c'est génial," dit l'homme en repoussant sa chaise et en sortant du restaurant sans payer. La serveuse a crié après lui, "Hé, et le projet de loi?" Elle haussa les épaules, marmonnant dans son souffle, "Pourquoi est-il si déformé?"

19. Google Translate's Italian translation of the "Restaurant" story: Un uomo andò in un ristorante e ordinò un hamburger, cucinato raro. Quando è arrivato, è stato bruciato per un croccante. La cameriera si fermò accanto al tavolo dell'uomo. "L'hamburger va bene?" Chiese lei. "Oh, è semplicemente fantastico," disse l'uomo, spingendo indietro la sedia e uscendo dal ristorante senza pagare. La cameriera gli urlò dietro, "Ehi, e il conto?" Lei scrollò le spalle, mormorando sottovoce, "Perché è così piegato?"

20. Google Translate's Chinese translation of the "Restaurant" story: 一名男子走进一家餐厅, 点了一个罕见的汉堡包. 当它到达时, 它被烧得脆脆. 女服务员停在男人的桌子旁边. "汉堡好吗" 她问. "哦, 这太好了," 那男人说, 推开椅子, 没有付钱就冲出餐厅. 女服务员大声喊道: "嘿, 账单呢?" 她耸了耸, 低声嘀咕道, "他为什么这么弯腰?"

21. For an in-depth discussion of the problems associated with Google Translate's lack of understanding, see D. R. Hofstadter, "The Shallowness of Google Translate," *The Atlantic*, Jan. 30, 2018.

22. D. R. Hofstadter, *Gödel, Escher, Bach: an Eternal Golden Braid* (New York: Basic Books, 1979), 603.

23. E. Davis and G. Marcus, "Commonsense Reasoning and Commonsense Knowledge in Artificial Intelligence," *Communications of the ACM* 58, no. 9 (2015): 92–103.

24. O. Vinyals et al., "Show and Tell: A Neural Image Caption Generator," in *Proceedings of the IEEE Conference on Computer Vision and Pattern Recognition* (2015), 3156–64; A. Karpathy and L. Fei-Fei, "Deep Visual-Semantic Alignments for Generating Image Descriptions," in *Proceedings of the IEEE Conference on Computer Vision and Pattern Recognition* (2015), 3128–37.

25. Figure 39 is a simplified version of the system described in Vinyals et al., "Show and Tell."

26. J. Markoff, "Researchers Announce Advance in Image-Recognition Software," *New York Times*, Nov. 17, 2014.

27. J. Walker, "Google's AI Can Now Caption Images Almost as Well as Humans," *Digital Journal*, Sept. 23, 2016, www.digitaljournal.com/tech-and-science/technology/google-s-ai-now-captions-images-with-94-accuracy/article/475547.

28. A. Linn, "Picture This: Microsoft Research Project Can Interpret, Caption Photos," *AI Blog*, May 28, 2015, blogs.microsoft.com/ai/picture-this-microsoft-research-project-can-interpret-caption-photos.

29. Microsoft CaptionBot, www.captionbot.ai.

13: Ask Me Anything

1. Transcript from www.chakoteya.net/NextGen/130.htm.

2. Quoted in F. Manjoo, "Where No Search Engine Has Gone Before," *Slate*, April 11, 2013, www.slate.com/articles/technology/technology/2013/04/google_has_a_single_towering_obsession_it_wants_to_build_the_star_trek_computer.html.

3. Quoted in C. Thompson, "What Is I.B.M.'s Watson?," *New York Times Magazine*, June 16, 2010.

4. Quoted in K. Johnson, "How 'Star Trek' Inspired Amazon's Alexa," *Venture Beat*, June 7, 2017, venturebeat.com/2017/06/07/how-star-trek-inspired-amazons-alexa.

5. *Wikipedia*, s.v. "Watson (computer)," accessed Dec. 16, 2018, en.wikipedia.org/wiki/Watson_(computer).

6. Thompson, "What Is I.B.M.'s Watson?"

7. A meme made popular on the television show *The Simpsons*.

8. K. Jennings, "The Go Champion, the Grandmaster, and Me," *Slate*, March 15, 2016, www.slate.com/articles/technology/technology/2016/03/google_s_alphago_defeated _go_champion_lee_sedol_ken_jennings_explains_what.html.

9. Quoted in D. Kawamoto, "Watson Wasn't Perfect: IBM Explains the 'Jeopardy!' Errors," Aol, accessed Dec. 16, 2018, www.aol.com/2011/02/17/the-watson-supercomputer -isnt-always-perfect-you-say-tomato.

10. J. C. Dvorak, "Was IBM's Watson a Publicity Stunt from the Start?," *PC Magazine*, Oct. 30, 2013, www.pcmag.com/article2/0,2817,2426521,00.asp.

11. M. J. Yuan, "Watson and Healthcare," IBM Developer website, April 12, 2011, www .ibm.com/developerworks/library/os-ind-watson/index.html.

12. "Artificial Intelligence Positioned to Be a Game-Changer," *60 Minutes*, Oct. 9, 2016, www.cbsnews.com/news/60-minutes-artificial-intelligence-charlie-rose-robot-sophia.

13. C. Ross and I. Swetlitz, "IBM Pitched Its Watson Supercomputer as a Revolution in Cancer Care. It's Nowhere Close," *Stat News*, Sept. 5, 2017, www.statnews.com/2017 /09/05/watson-ibm-cancer.

14. P. Rajpurkar et al., "SQuAD: 100,000+ Questions for Machine Comprehension of Text," in *Proceedings of the 2016 Conference on Empirical Methods in Natural Language Processing* (2016), 2383–92.

15. Ibid.

16. A. Linn, "Microsoft Creates AI That Can Read a Document and Answer Questions About It as Well as a Person," *AI Blog*, Microsoft, Jan. 15, 2018, blogs.microsoft.com /ai/microsoft-creates-ai-can-read-document-answer-questions-well-person.

17. Quoted in "AI Beats Humans at Reading Comprehension for the First Time," Technology.org, Jan. 17, 2018, www.technology.org/2018/01/17/ai-beats-humans -at-reading-comprehension-for-the-first-time.

18. D. Harwell, "AI Models Beat Humans at Reading Comprehension, but They've Still Got a Ways to Go," *Washington Post*, Jan. 16, 2018.

19. P. Clark et al., "Think You Have Solved Question Answering? Try ARC, the AI2 Reasoning Challenge," arXiv:1803.05457 (2018).

20. Ibid.

21. ARC Dataset Leaderboard, Allen Institute for Artificial Intelligence, accessed Dec. 17, 2018, leaderboard.allenai.org/arc/submissions/public.

22. All of the examples in this section are from E. Davis, L. Morgenstern, and C. Ortiz, "The Winograd Schema Challenge," accessed Dec. 17, 2018, cs.nyu.edu/faculty/davise /papers/WS.html.

23. T. Winograd, *Understanding Natural Language* (New York: Academic Press, 1972).

24. H. J. Levesque, E. Davis, and L. Morgenstern, "The Winograd Schema Challenge," in *AAAI Spring Symposium: Logical Formalizations of Commonsense Reasoning* (American Association for Artificial Intelligence, 2011), 47.

25. T. H. Trinh and Q. V. Le, "A Simple Method for Commonsense Reasoning," arXiv:1806.02847 (2018).

26. Quoted in K. Bailey, "Conversational AI and the Road Ahead," *Tech Crunch*, Feb. 25, 2017, techcrunch.com/2017/02/25/conversational-ai-and-the-road-ahead.

27. H. Chen et al., "Attacking Visual Language Grounding with Adversarial Examples: A Case Study on Neural Image Captioning," in *Proceedings of the 56th Annual Meeting of the Association for Computational Linguistics*, vol. 1, *Long Papers* (2018), 2587–97.

28. N. Carlini and D. Wagner, "Audio Adversarial Examples: Targeted Attacks on Speech-to-Text," in *Proceedings of the First Deep Learning and Security Workshop* (2018).

29. R. Jia and P. Liang, "Adversarial Examples for Evaluating Reading Comprehension Systems," in *Proceedings of the 2017 Conference on Empirical Methods in Natural Language Processing* (2017).

30. C. D. Manning, "Last Words: Computational Linguistics and Deep Learning," *Nautilus*, April 2017.

14: On Understanding

1. G.-C. Rota, "In Memoriam of Stan Ulam: The Barrier of Meaning," *Physica D Nonlinear Phenomena* 22 (1986): 1–3.

2. At one lecture I gave on this topic, a student asked, "Why does an AI system need to have a humanlike understanding? Why can't we accept AI with a different kind of understanding?" Beyond the fact that I don't have any idea what a "different kind of understanding" would mean, my point is that if AI systems are to interact with humans in the world, they need to understand the situations they encounter in essentially the same way humans do.

3. The term *core knowledge* has been used most prominently by the psychologist Elizabeth Spelke and her collaborators; for example, see E. S. Spelke and K. D. Kinzler, "Core Knowledge," *Developmental Science* 10, no. 1 (2007): 89–96. Many other cognitive scientists have discussed similar ideas.

4. Psychologists use the term *intuitive* because this basic knowledge is so ingrained in our minds from an early age; this knowledge becomes self-evident to us, and for the most part it remains subconscious. Numerous psychologists have shown that there are aspects of typical human intuitive beliefs about physics, probability, and other areas that are actually erroneous. See, for example, A. Tversky and D. Kahneman, "Judgment Under Uncertainty: Heuristics and Biases," *Science* 185, no. 4157 (1974): 1124–31; and B. Shanon, "Aristotelianism, Newtonianism, and the Physics of the Layman," *Perception* 5, no. 2 (1976): 241–43.

5. Lawrence Barsalou gives a detailed argument for such mental simulations in L. W. Barsalou, "Perceptual Symbol Systems," *Behavioral and Brain Sciences* 22 (1999): 577–660.

6. Douglas Hofstadter points out that when one encounters (or remembers, or reads about, or imagines) a situation, the representation of the situation in one's mind includes a "halo" of possible variations on that situation that he calls an "implicit counterfactual sphere," which includes "the things that never were but that we cannot help seeing anyway." D. R. Hofstadter, *Metamagical Themas* (New York: Basic Books, 1985), 247.

7. L. W. Barsalou, "Grounded Cognition," *Annual Review of Psychology* 59 (2008): 617–45.

8. L. W. Barsalou, "Situated Simulation in the Human Conceptual System," *Language and Cognitive Processes* 18, no. 5–6 (2003): 513–62.

9. A.E.M. Underwood, "Metaphors," *Grammarly* (blog), accessed Dec. 17, 2018, www.grammarly.com/blog/metaphor.

10. G. Lakoff and M. Johnson, *Metaphors We Live By* (Chicago: University of Chicago Press, 1980).

11. L. E. Williams and J. A. Bargh, "Experiencing Physical Warmth Promotes Interpersonal Warmth," *Science* 322, no. 5901 (2008): 606–607.

12. C. B. Zhong and G. J. Leonardelli, "Cold and Lonely: Does Social Exclusion Literally Feel Cold?," *Psychological Science* 19, no. 9 (2008): 838–42.

13. D. R. Hofstadter, *I Am a Strange Loop* (New York: Basic Books, 2007). The quotation is from the front book flap. My description also echoes ideas proposed by the philosopher Daniel Dennett in his book *Consciousness Explained* (New York: Little, Brown, 1991).

14. This kind of "linguistic productivity" is discussed in D. Hofstadter and E. Sander, *Surfaces and Essences: Analogy as the Fuel and Fire of Thinking* (New York: Basic Books, 2013), 129, and in A. M. Zwicky and G. K. Pullum, "Plain Morphology and Expressive Morphology," in *Annual Meeting of the Berkeley Linguistics Society* (1987), 13:330–40.

15. I borrowed this argument from an actual legal case. See "Blogs as Graffiti? Using Analogy and Metaphor in Case Law," *IdeaBlawg*, March 17, 2012, www.ideablawg.ca /blog/2012/3/17/blogs-as-graffiti-using-analogy-and-metaphor-in-case-law.html.

16. D. R. Hofstadter, "Analogy as the Core of Cognition," Presidential Lecture, Stanford University (2009), accessed Dec. 18, 2018, www.youtube.com/watch?v =n8m7lFQ3njk.

17. Hofstadter and Sander, *Surfaces and Essences*, 3.

18. M. Minsky, "Decentralized Minds," *Behavioral and Brain Sciences* 3, no. 3 (1980): 439–40.

15: Knowledge, Abstraction, and Analogy in Artificial Intelligence

1. D. B. Lenat and J. S. Brown, "Why AM and EURISKO Appear to Work," *Artificial Intelligence* 23, no. 3 (1984): 269–94.

2. These examples are from C. Metz, "One Genius' Lonely Crusade to Teach a Computer Common Sense," *Wired*, March 24, 2016, www.wired.com/2016/03/doug-lenat -artificial-intelligence-common-sense-engine, and D. Lenat, "Computers Versus Common Sense," Google Talks Archive, accessed Dec. 18, 2018, www.youtube.com /watch?v=gAtn-4fhuWA.

3. Lenat notes that the company is increasingly able to automate the process of obtaining new assertions (presumably by mining the web). From D. Lenat, "50 Shades of Symbolic Representation and Reasoning," CMU Distinguished Lecture Series, accessed Dec. 18, 2018, www.youtube.com/watch?v=4mv0nCS2mik.

4. Ibid.

5. A detailed, nontechnical description of the Cyc project is given in chapter 4 of H. R. Ekbia, *Artificial Dreams: The Quest for Non-biological Intelligence* (Cambridge, U.K.: Cambridge University Press, 2008).

6. Lucid company's webpage: lucid.ai.

7. P. Domingos, *The Master Algorithm* (New York: Basic Books, 2015), 35.

8. From "The Myth of AI: A Conversation with Jaron Lanier," *Edge*, Nov. 14, 2014, www .edge.org/conversation/jaron_lanier-the-myth-of-ai.

9. For example, see N. Watters et al., "Visual Interaction Networks," *Advances in Neural Information Processing Systems* 30 (2017): 4539–47; T. D. Ullman et al., "Mind Games: Game Engines as an Architecture for Intuitive Physics," *Trends in Cognitive Sciences* 21, no. 9 (2017): 649–65; and K. Kansky et al., "Schema Networks: Zero-Shot Transfer with a Generative Causal Model of Intuitive Physics," in *Proceedings of the International Conference on Machine Learning* (2017), 1809–18.

10. J. Pearl, "Theoretical Impediments to Machine Learning with Seven Sparks from the Causal Revolution," in *Proceedings of the Eleventh ACM International Conference on Web Search and Data Mining* (2018), 3. For a more in-depth discussion of causal reasoning in AI, see J. Pearl and D. Mackenzie, *The Book of Why: The New Science of Cause and Effect* (New York: Basic Books, 2018).

11. For an insightful discussion on what is missing in deep learning, see G. Marcus, "Deep Learning: A Critical Appraisal," arXiv:1801.00631 (2018).

12. DARPA Fiscal Year 2019 Budget Estimates, Feb. 2018, accessed Dec. 18, 2018, www.darpa.mil/attachments/DARPAFY19PresidentsBudgetRequest.pdf.

13. English version: M. Bongard, *Pattern Recognition* (New York: Spartan Books, 1970).

14. All of the Bongard-problem images I give here are from Harry Foundalis's Index of Bongard Problems website, www.foundalis.com/res/bps/bpidx.htm, which gives Bongard's one hundred problems as well as many problems created by other people.

15. R. M. French, *The Subtlety of Sameness* (Cambridge, Mass.: MIT Press, 1995).

16. One particularly interesting program that attempted to solve Bongard problems was created by Harry Foundalis when he was a graduate student in Douglas Hofstadter's research group at Indiana University. Foundalis explicitly declared that he was building not a "Bongard problem solver" but a "cognitive architecture inspired by Bongard's problems." The program was inspired by humanlike perception at every level, from low-level vision all the way to abstraction and analogy, very much in the spirit of Bongard's intentions, though it was successful in solving only a small number of Bongard's problems. See H. E. Foundalis, "Phaeaco: A Cognitive Architecture Inspired by Bongard's Problems" (PhD diss., Indiana University, 2006), www.foundalis.com/res/Foundalis_dissertation.pdf. Foundalis maintains an extensive website related to his work on Bongard problems: www.foundalis.com/res/diss_research.html.

17. S. Stabinger, A. Rodríguez-Sánchez, and J. Piater, "25 Years of CNNs: Can We Compare to Human Abstraction Capabilities?," in *Proceedings of the International Conference on Artificial Neural Networks* (2016), 380–87. A related study with similar results was reported in J. Kim, M. Ricci, and T. Serre, "Not-So-CLEVR: Visual Relations Strain Feedforward Neural Networks," *Interface Focus* 8, no. 4 (2018): 2018.0011.

18. When I say "most people," I am referring to the results of surveys I gave to people as part of my dissertation work. See M. Mitchell, *Analogy-Making as Perception* (Cambridge, Mass.: MIT Press, 1993).

19. Hofstadter coined the term *conceptual slippage* in his discussion of Bongard problems in chapter 19 of D. R. Hofstadter, *Gödel, Escher, Bach: an Eternal Golden Braid* (New York: Basic Books, 1979).

20. Ibid., 349–51.

21. A detailed description of Copycat is given in chapter 5 of D. R. Hofstadter and the Fluid Analogies Research Group, *Fluid Concepts and Creative Analogies: Computer Models of the Fundamental Mechanisms of Thought* (New York: Basic Books, 1995). An even more detailed description is given in the book based on my dissertation: Mitchell, *Analogy-Making as Perception*.

22. J. Marshall, "A Self-Watching Model of Analogy-Making and Perception," *Journal of Experimental and Theoretical Artificial Intelligence* 18, no. 3 (2006): 267–307.

23. Several of these programs are described in Hofstadter and the Fluid Analogies Research Group, *Fluid Concepts and Creative Analogies*.

24. A. Karpathy, "The State of Computer Vision and AI: We Are Really, Really Far Away," Andrej Karpathy blog, Oct. 22, 2012, karpathy.github.io/2012/10/22/state-of-computer-vision.

25. See *Stanford Encyclopedia of Philosophy*, s.v. "Dualism," plato.stanford.edu/entries/dualism/.

26. For a cogent philosophical discussion of the embodiment hypothesis in cognitive science, see A. Clark, *Being There: Putting Brain, Body, and World Together Again* (Cambridge, Mass.: MIT Press, 1996).

16: Questions, Answers, and Speculations

1. "Automated Vehicles for Safety," National Highway Traffic Safety Administration website, www.nhtsa.gov/technology-innovation/automated-vehicles-safety#issue-road -self-driving.

2. "Vehicle Cybersecurity: DOT and Industry Have Efforts Under Way, but DOT Needs to Define Its Role in Responding to a Real-World Attack," General Accounting Office, March 2016, accessed Dec. 18, 2018, www.gao.gov/assets/680/676064 .pdf.

3. Quoted in J. Crosbie, "Ford's Self-Driving Cars Will Live Inside Urban 'Geofences,'" *Inverse*, March 13, 2017, www.inverse.com/article/28876-ford-self-driving-cars-geo fences-ride-sharing.

4. Quoted in J. Kahn, "To Get Ready for Robot Driving, Some Want to Reprogram Pedestrians," *Bloomberg*, Aug. 16, 2018, www.bloomberg.com/news/articles/2018-08 -16/to-get-ready-for-robot-driving-some-want-to-reprogram-pedestrians.

5. "Artificial Intelligence, Automation, and the Economy," Executive Office of the President, Dec. 2016, www.whitehouse.gov/sites/whitehouse.gov/files/images/EMBAR GOED%20AI%20Economy%20Report.pdf.

6. This harks back to what Alan Turing called "Lady Lovelace's objection," named for Lady Ada Lovelace, a British mathematician and writer who worked with Charles Babbage on developing the Analytical Engine, a nineteenth-century proposal for a (never completed) programmable computer. Turing quotes from Lady Lovelace's writings: "The Analytical Engine has no pretensions to *originate* anything. It can do *whatever we know how to order it* to perform." A. M. Turing, "Computing Machinery and Intelligence," *Mind* 59, no. 236 (1950): 433–60.

7. Karl Sims website, accessed Dec. 18, 2018, www.karlsims.com.

8. D. Cope, *Virtual Music: Computer Synthesis of Musical Style* (Cambridge, Mass.: MIT Press, 2004).

9. Quoted in G. Johnson, "Undiscovered Bach? No, a Computer Wrote It," *New York Times*, Nov. 11, 1997.

10. M. A. Boden, "Computer Models of Creativity," *AI Magazine* 30, no. 3 (2009): 23–34.

11. J. Gottschall, "The Rise of Storytelling Machines," in *What to Think About Machines That Think*, ed. J. Brockman (New York: Harper Perennial, 2015), 179–80.

12. From "Creating Human-Level AI: How and When?," video lecture, Future of Life Institute, Feb. 9, 2017, www.youtube.com/watch?v=V0aXMTpZTfc.

13. A. Karpathy, "The State of Computer Vision and AI: We Are Really, Really Far Away," Andrej Karpathy blog, Oct. 22, 2012, karpathy.github.io/2012/10/22/state -of-computer-vision.

14. C. L. Evans, *Broad Band: The Untold Story of the Women Who Made the Internet* (New York: Portfolio/Penguin, 2018), 24.

15. M. Campbell-Kelly et al., *Computer: A History of the Information Machine*, 3rd ed. (New York: Routledge, 2018), 80.

16. Quoted in K. Anderson, "Enthusiasts and Skeptics Debate Artificial Intelligence," *Vanity Fair*, Nov. 26, 2014.

17. See O. Etzioni, "No, the Experts Don't Think Superintelligent AI Is a Threat to Humanity," *Technology Review*, Sept. 20, 2016, www.technologyreview.com/s/602410/no-the-experts-dont-think-superintelligent-ai-is-a-threat-to-humanity; and V. C. Müller and N. Bostrom, "Future Progress in Artificial Intelligence: A Survey of Expert Opinion," in *Fundamental Issues of Artificial Intelligence* (Basel, Switzerland: Springer, 2016), 555–72.

18. N. Bostrom, "How Long Before Superintelligence?," *International Journal of Future Studies* 2 (1998).

19. D. R. Hofstadter, *Gödel, Escher, Bach: an Eternal Golden Braid* (New York: Basic Books, 1979), 677–78.

20. From "The Myth of AI: A Conversation with Jaron Lanier," *Edge*, Nov. 14, 2014, www.edge.org/conversation/jaron_lanier-the-myth-of-ai.

21. P. Domingos, *The Master Algorithm* (New York: Basic Books, 2015), 285–86.

22. From "Panel: Progress in AI: Myths, Realities, and Aspirations," Microsoft Research video, accessed Dec. 18, 2018, www.youtube.com/watch?v=1wPFEj1ZHRQ&feature=youtu.be.

23. R. Brooks, "The Origins of 'Artificial Intelligence,'" Rodney Brooks's blog, April 27, 2018, rodneybrooks.com/forai-the-origins-of-artificial-intelligence.

Acknowledgments

This book owes its existence to Douglas Hofstadter. Doug's writings were what attracted me to AI in the first place, and his ideas and mentorship guided my PhD studies. More recently, Doug invited me to the meeting at Google that sparked the idea for this book, and even more recently, he read every chapter of the manuscript, filling the pages with insightful comments that greatly improved the final version. I'm very thankful for Doug's ideas, his books and articles, his support of my work, and above all, his friendship.

I am grateful to several other friends and family members who generously read and perceptively commented on every chapter: Jim Levenick, Jim Marshall, Russ McBride, Jack Mitchell, Norma Mitchell, Kendall Springer, and Chris Wood. Many thanks also to the following people for answering questions, translating passages, and offering other types of assistance: Jeff Clune, Richard Danzig, Bob French, Garrett Kenyon, Jeff Kephart, Blake LeBaron, Sheng Lundquist, Dana Moser, David Moser, and Francesca Parmeggiani.

Much gratitude to Eric Chinski at Farrar, Straus and Giroux for his encouragement and ever-astute contributions on all aspects of this project; to Laird Gallagher for the many thoughtful suggestions that helped turn a rough manuscript into a finished text; and to the rest of the team at FSG, especially Julia Ringo, Ingrid Sterner, Rebecca Caine, Richard Oriolo, Deborah Ghim, and Brian Gittis, for all their great work. Many thanks also to my agent, Esther Newberg, for helping to make this book a reality.

I owe much appreciation to my husband, Kendall Springer, for his constant love and enthusiastic support, plus his patient tolerance of my crazy work habits. My sons, Jacob and Nicholas Springer, have been a wonderful inspiration over the years with their remarkable questions, curiosity, and common sense. This book is dedicated to my parents, Jack and Norma Mitchell, who have provided me with unlimited encouragement and love throughout my life. In a world full of machines, I am very fortunate to be surrounded by such sage and loving humans.

Index

25   Figure 1: Drawing of neuron adapted from C. Ling, M. L. Hendrickson, and R. E. Kalil, "Resolving the Detailed Structure of Cortical and Thalamic Neurons in the Adult Rat Brain with Refined Biotinylated Dextran Amine Labeling," *PLOS ONE* 7, no. 11 (2012), e45886. Image licensed under Creative Commons Attribution 4.0 International license (creativecommons.org/licenses/by/4.0/).

26   Figure 2: Handwritten characters image by Josef Steppan, commons.wikimedia .org/wiki/File:MnistExamples.png. Image licensed under Creative Commons Attribution-ShareAlike 4.0 International license (creativecommons.org/licenses /by-sa/4.0/deed.en).

27   Figure 3: Author.

36   Figure 4: Author.

56   Figure 5: Author.

68   Figure 6: media.defense.gov/2015/May/15/2001047923/-1/-1/0/150506-F-BD468-053 .JPG, accessed Dec. 4, 2018 (public domain).

70   Figure 7: Author.

72   Figure 8: Author.

73   Figure 9: Author.

74   Figure 10: Author.

75   Figure 11: Author.

76   Figure 12: Author.

101   Figure 13: Author.

103   Figure 14: From twitter.com/amywebb/status/841292068488118273, accessed Dec. 7, 2018. Reprinted by permission of Amy Webb.

105   Figure 15: www.nps.gov/yell/learn/nature/osprey.htm (public domain); www.fs.usda.gov/Internet/FSE_MEDIA/stelprdb5371680.jpg (public domain).

106   Figure 16: From twitter.com/jackyalcine/status/615329515909156865, accessed Dec. 7, 2018. Reprinted by permission of Jacky Alcine.

107   Figure 17: From www.flickr.com/photos/jozjozjoz/352910684, accessed Dec. 7, 2018. Reprinted by permission of Joz Wang of jozjozjoz.com.

111   Figure 18: From C. Szegedy et al., "Intriguing Properties of Neural Networks," in *Proceedings of the International Conference on Learning Representations* (2014). Reprinted by permission of Christian Szegedy.

112   Figure 19: From A. Nguyen, J. Yosinski, and J. Clune, "Deep Neural Networks Are Easily Fooled: High Confidence Predictions for Unrecognizable Images," in *Proceedings of the IEEE Conference on Computer Vision and Pattern Recognition* (2015), 427–36. Reprinted by permission of the authors.

113   Figure 20: Figure adapted from M. Sharif et al., "Accessorize to a Crime: Real and Stealthy Attacks on State-of-the-Art Face Recognition," in *Proceedings of the 2016 ACM SIGSAC Conference on Computer and Communications Security* (2016), 1528–40. Reprinted by permission of the authors. Milla Jovovich photograph is from commons.wikimedia.org/wiki/File:Milla_Jovovich.png, by Georges Biard, licensed under Creative Commons Attribution-Share Alike 3.0 Unported license (creative-commons.org/licenses/by-sa/3.0/deed.en).

115   Figure 21: Author.

134   Figure 22: From www.cs.cmu.edu/~robosoccer/image-gallery/legged/2003/aibo-with-ball12.jpg. Reprinted by permission of Manuela Veloso.

137   Figure 23: Author.

139   Figure 24: Author.

140   Figure 25: Author.

141   Figure 26: Author.

146   Figure 27: Author.

148   Figure 28: Author.

151   Figure 29: Author.

153   Figure 30: Author.

161   Figure 31: Author.

185   Figure 32: Author.

186   Figure 33: Author.

189   Figure 34: Author.

191   Figure 35: Author.

193   Figure 36: Author.

194   Figure 37: From T. Mikolov et al., "Distributed Representations of Words and Phrases and Their Compositionality," in *Advances in Neural Information Processing Systems* (2013), 3111–19. Reprinted by permission of Tomas Mikolov.

200   Figure 38: Author.

209   Figure 39: Author. Photograph is from the Microsoft COCO data set: cocodataset.org.

209   Figure 40: Photograph and captions are from the Microsoft COCO data set: coco
       dataset.org.

210   Figure 41: Photographs and captions are from nic.droppages.com. Reprinted by
       permission of Oriol Vinyals.

211   Figure 42: Top row: Photographs and captions from O. Vinyals et al., "Show and
       Tell: A Neural Image Caption Generator," in *Proceedings of the IEEE Conference on
       Computer Vision and Pattern Recognition* (2015), 3156–64. Reprinted by permission of
       Oriol Vinyals. Bottom row, left: Road-Tech Safety Services. Reprinted by permis-
       sion of Ben Jeffrey. Bottom row, right: Nikoretro, https://www.flickr.com/photos
       /bellatrix6/4727507323/in/album-72057594083648059. Licensed under Creative Com-
       mons Attribution-ShareAlike 2.0 Generic license: https://creativecommons.org
       /licenses/by-sa/2.0/. Bottom row captions are from captionbot.ai.

229   Figure 43: Photographs and captions from H. Chen et al., "Attacking Visual Lan-
       guage Grounding with Adversarial Examples: A Case Study on Neural Image Cap-
       tioning," in *Proceedings of the 56th Annual Meeting of the Association for Computational
       Linguistics*, vol. 1, *Long Papers* (2018), 2587–97. Reprinted with permission of Hongge
       Chen and the Association for Computational Linguistics.

236   Figure 44: Dorothy Alexander / Alamy Stock Photo.

252   Figure 45: From www.foundalis.com/res/bps/bpidx.htm. Original images are
       from M. Bongard, *Pattern Recognition* (New York: Spartan Books, 1970).

253   Figure 46: From www.foundalis.com/res/bps/bpidx.htm. Original images are
       from M. Bongard, *Pattern Recognition* (New York: Spartan Books, 1970).

255   Figure 47: Author.

262   Figure 48: Photographs taken by author.

263   Figure 49: www.nps.gov/dena/planyourvisit/pets.htm (public domain); pxhere
       .com/en/photo/1394259 (public domain); Peter Titmuss / Alamy Stock Photo; Thang
       Nguyen, www.flickr.com/photos/70209763@N00/399996115, licensed under Cre-
       ative Commons Attribution-ShareAlike 2.0 Generic license (creativecommons.org
       /licenses/by-sa/2.0/).

264   Figure 50: P. Souza, *Obama: An Intimate Portrait* (New York: Little, Brown, 2018), 102
       (public domain).